Miss Leavitt's Stars

PUBLISHED TITLES IN THE GREAT DISCOVERIES SERIES

David Foster Wallace
Everything and More: A Compact History of ∞

Sherwin B. Nuland
The Doctors' Plague: Germs, Childbed Fever, and the Strange Story of Ignác Semmelweis

Michio Kaku
Einstein's Cosmos: How Albert Einstein's Vision Transformed Our Understanding of Space and Time

Barbara Goldsmith
Obsessive Genius: The Inner World of Marie Curie

Rebecca Goldstein
Incompleteness: The Proof and Paradox of Kurt Gödel

Madison Smartt Bell
Lavoisier in the Year One: The Birth of a New Science in an Age of Revolution

George Johnson
Miss Leavitt's Stars: The Untold Story of the Woman Who Discovered How to Measure the Universe

David Leavitt
The Man Who Knew Too Much: Alan Turing and the Invention of the Computer

William T. Vollman
Uncentering the Earth: Copernicus and The Revolutions of the Heavenly Spheres

FORTHCOMING TITLES

Richard Reeves on Rutherford and the Atom

Daniel Mendelsohn on Archimedes and the Science of the Ancient Greeks

David Quammen on Darwin and Evolution

Lawrence Krauss on the Science of Richard Feynman

General Editors: Edwin Barber and Jesse Cohen

BY GEORGE JOHNSON

Architects of Fear: Conspiracy Theories and Paranoia in American Politics

Machinery of the Mind: Inside the New Science of Artificial Intelligence

In the Palaces of Memory: How We Build the Worlds Inside Our Heads

Fire in the Mind: Science, Faith, and the Search for Order

Strange Beauty: Murray Gell-Mann and the Revolution in Twentieth-Century Physics

A Shortcut Through Time: The Path to the Quantum Computer

Miss Leavitt's Stars: The Untold Story of the Woman Who Discovered How to Measure the Universe

GEORGE JOHNSON

Miss Leavitt's Stars

The Untold Story of the Woman Who Discovered How to Measure the Universe

ATLAS BOOKS

W. W. NORTON & COMPANY
NEW YORK • LONDON

Printed in the United States of America
First published as a Norton paperback 2006

Manufacturing by R. R. Donnelley, Harrisonburg Division
Book design by Chris Welch
Production manager: Julia Druskin

Library of Congress Cataloging-in-Publication Data

Johnson, George, date.
Miss Leavitt's stars : the untold story of the woman who discovered how to measure the universe / George Johnson.— 1st ed.
p. cm. — (Great discoveries)
Includes bibliographical references and index.
ISBN 0-393-05128-5
1. Astrometry—History. 2. Leavitt, Henrietta Swan, 1868–1921. 3. Astronomical photometry. 4. Astronomy—United States—History—20th century.
I. Title. II. Series.

QB807.J64 2005
522'.09'04—dc22

2005002823

ISBN-13: 978-0-393-32856-1 pbk.
ISBN-10: 0-393-32856-2 pbk.

Atlas Books, 10 E. 53rd St., 35th Fl., New York NY 10022

W. W. Norton & Company, Inc., 500 Fifth Avenue, New York, N.Y. 10110
www.wwnorton.com

W. W. Norton & Company Ltd., Castle House, 75/76 Wells Street, London W1T 3QT

1 2 3 4 5 6 7 8 9 0

For my mother, Dorris M. Johnson

Her columns grew longer, and if she squinted at them, the confetti of inklings began to resemble a skyful of stars. She had time to let her mind wander. The Magi's search for Bethlehem; the music of Milton's crystal spheres . . . they could all be reduced to these numbers. There was actually no need to squint and pretend that the digits were the stars. They were, by themselves, wildly alive, fact and symbol of the vast, cool distances in which one located the light of different worlds.

—THOMAS MALLON, *Two Moons*

Then, by means of the instrument at hand, they travelled together from the earth to Uranus, and the mysterious outskirts of the solar system; from the solar system to a star in the Swan, the nearest fixed star in the northern sky; from the star in the Swan to remoter stars; thence to the remotest visible; till the ghastly chasm which they had bridged by a fragile line of sight was realized. . . .

—THOMAS HARDY, *Two on a Tower*

Contents

Preface

Henrietta Swan Leavitt deserves a proper biography. She will probably never get one, so faint is the trail she left behind. No personal diaries, no boxes of letters, no scientific memoirs—she wasn't one to brag. She rates no more than a few paragraphs in the biographical reference books, the *Who's Who* of this or that, a footnote or a boxed sidebar in the Astronomy 101 texts.

I had intended to use her as nothing more than a device, a way to get into the story of how, in the 1920s, people learned that there is more to the universe than just the Milky Way. The discovery she made in her menial position at the Harvard Observatory was the turning point. I figured I would make that turn in the first chapter and then move on.

But Henrietta Leavitt refused to exit on cue from the story. I couldn't get this woman out of my mind. Why, given the remarkable, completely unexpected nature of her observation, didn't she push beyond it, working shoulder to shoulder with the great Harlow Shapley and Edwin Hubble, as they grabbed onto Henrietta's law and ran with it light-years across the sky?

And so, just when I thought the book was almost done, I found myself going back to the beginning, searching her genealogy, scouring the records as they had apparently not been scoured before. Scrap by scrap, the apparition in the canned biographical sketches was replaced by a human being—not solid enough, perhaps, to star in her own biography but someone with a story to tell.

PROLOGUE

The Village in the Canyon

The village was hidden at the bottom of a deep chasm with sides so steep and slick that no one had ever climbed them. All that could be seen overhead was a narrow band of sky.

Looking down the canyon, one could spot a hill in the distance. How distant no one knew, for it was separated from the village by an impassable expanse. Beyond the hill, and even more inaccessible, was a faraway mountain, the edge of the knowable world.

The villagers had noticed that when they walked the width of their canyon, from one wall to the other, the top of the hill appeared to shift ever so slightly against the backdrop of the mountain, moving from one side of the peak to the other. No one believed the hill really moved, but they enjoyed the illusion.

One day a particularly observant citizen noticed an interesting subtlety. When he wandered up the canyon a ways, and then traversed its width from wall to wall, the hilltop still appeared to move, but much more slowly. And if he walked even farther and repeated the experiment, the movement was smaller still. Venture far enough, he discovered, and there was no perceptible shift at all.

He made an entry in his notebook: The amount the hill moves depends on its distance from the observer. He had discovered the phenomenon called parallax.

Returning to the village, where the hill was closest and the effect was most pronounced, he measured the separation between the canyon walls and began to sketch out a picture:

The width of the canyon and the line of sight from each canyon wall to the center of the hilltop formed an imaginary triangle. Using a surveying instrument, he could measure the two angles at the triangle's base. Then, using what he had learned in school about the rules of trigonometry, he calculated the height of the triangle—the distance from the center of the baseline to the apex, from the village straight across the impassable badlands to the hilltop. By this measure, it was ten canyon widths

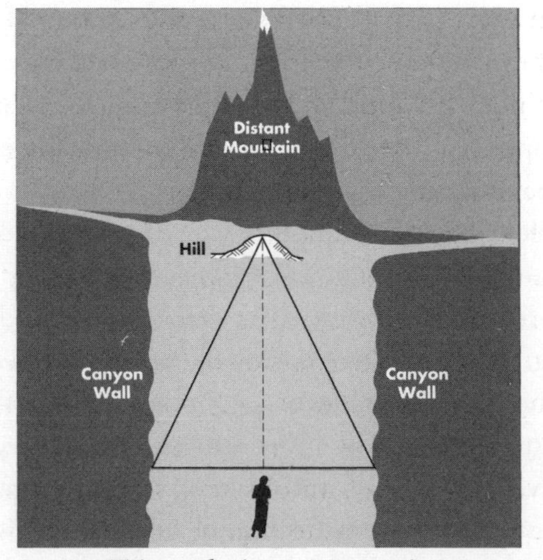

Triangulating a Mountain

away. It was still impossible to walk there, but it was comforting to know the distance. The world seemed a little tamer.

In coming years, the villagers began to build stone lookout towers, climbing high enough to observe that beyond the hill was another one. Using the same measuring technique, they showed that the second hill was fifteen canyon widths away. Behind it was a third hill at twenty-five canyon widths. But beyond that the method failed. The hills were so far that they couldn't be made to shift at all. The villagers' canyon was not wide enough, their triangles were too small.

Most tantalizing of all was the dark immovable mountain on the horizon. It seemed to lie at infinity, a distance so great as to be immeasurable. A few villagers imagined launching an expedition, traversing the dangerous badlands between the village and the first hill and then forging beyond that, walking day after day until they had reached the mountain. But none was so foolish or so brave. The more imaginative citizens could envision what it might be like to rise up out of the canyon, where they would be free to move so far left and right that the mountain itself would shift against some even more distant backdrop. Then they could determine how long the journey would be. But that was just fantasy.

In theory, there was another way of indirectly determining the mountain's distance: things looked smaller the farther away they were. Something twice as distant would appear half as high. If this rule applied outside the immediate area of the village—and why would it not?—there would be a means of measuring to the mountain.

One very clear day, an observer standing on the highest tower decided to try this technique. She had noticed long ago that there was vegetation on the nearest hilltop—appearing in the distance like miniature versions of the local flora, the spiny green bushes that covered the canyon floor. On this day she noticed that if she squinted her eyes, she could also make out, just barely, the tiniest fringe of green along the mountaintop, comforting evidence that this far-off land might not be unlike her own.

It took only a few moments for her to appreciate the larger implication. Working carefully, she measured the apparent height of the shrunken shrubs on the nearest hill. The even tinier fringe of green on the mountain appeared to be some ten times smaller. And so, it seemed, the mountain itself must be ten times farther than the hilltop—a full one hundred canyon widths away.

That was far but not infinite. Walking there might take a week or two—if someone could find a way to cross the impassable divide. Encouraged by her discovery, a party of volunteers set out for the mountain. From the highest towers the citizens watched as the explorers eventually found a way across the badlands, to the first hilltop and on past the second and the third, until in a few days they were out of sight.

After a week had passed, the villagers went back to the towers to watch for the expedition's return. Two weeks later they looked again. Months went by, then a year before they stopped waiting.

Finally one day, a single member of the party staggered back home. The vegetation on the mountaintop, he reported, was very different from anything in the canyon. Towering trees rose ten times higher than any plant he had ever seen. He had climbed to the top of one of these behemoths and thought he could barely make out the tiny twinkling lights back home. He knew by then that while the logic of the measurement had been sound, the villagers had been misled by the limits of their imagination. Since the strange trees were ten times higher than the familiar bushes, the mountain was another ten times farther than had been reckoned, a full one thousand canyon widths away. . . .

2

In Robert Heinlein's novel *Time for the Stars*, a charitable organization called the Long Range Foundation recruits pairs of identical twins for a mission to colonize space. The Earth, buckling under an exploding population, has already spilled over to all the planets of the solar system. Now it must look farther, to planets orbiting stars so distant that news of their discovery, traveling at light-speed, would take years to reach home.

That is where the twins come in. Scientists, according to the story, have discovered that many twins are telepathic, and that the signals they mentally exchange are not weighed down by the restrictions imposed on electromagnetic waves. The communication is instantaneous—faster than light—no matter how far apart the twins may be. With one twin on a spaceship and the other on Earth, they can converse instantly across light-years.

I read this story, now out of print, in junior high school and had forgotten all but the bare bones of the plot, which revolves around the relativistic effects of traveling near light-speed. Time slows down so Tom, the twin on the starship, ages only a fraction as quickly as Pat, his counterpart back home. He is an old man when Tom returns and marries Pat's granddaughter, with whom he has been flirting telepathically.

What stuck in my mind all those years was not the rather obvious Einsteinian plot contrivances but a brilliant scene toward the end of the book: Tom is looking at the sky from a hospitable planet orbiting Tau Ceti, a star system eleven light-years from Earth. The constellations he sees are recognizable but slightly distorted from how they appear back home. He can make out the Big Dipper, "looking a little more battered than it does from Earth," and he finds Orion, the great hunter, though his dog, Sirius, is stretched out of whack. The understated climax comes when Tom discovers, from where he now stands, that the constellation Boötes has acquired a new star, yellow-white and of about the second magnitude. It takes him a moment to realize that he is looking back at the sun.

I don't know why that scene threw me for a loop. Maybe it should have been obvious even for an eighth grader how arbitrary the constellations are, not just the names taken from classical mythology (does anyone really think Ursa Major looks like a giant bear?) but the actual shapes. They are an accident of how the stars happen to line up from one among an infinity of possible points of view. Though Orion's dog appears to follow faithfully at his heels, the constellation's principal star, Sirius, 8.6 light-years from Earth, is nowhere near the hunter's belt, whose stars are approximately 1,500 light-years away.

For that matter Orion's belt is nowhere near his shoulders and the shoulders nowhere near his knees. Even the belt is an accident of perspective. From other parts of space these three stars would form a triangle or their order would be reversed. Viewed from just the proper angle, all three of them would merge into a single light, an illusory triple star. Travel to the center of the space the three stars enclose and each would appear in a far-flung corner of the sky.

3

For anyone raised on science fiction or the enthusiastic promises of the Kennedy era, the space program has turned out to be a dud. Who would have guessed that several decades later, after a few trips to the moon (about 250,000 miles or fifty round trips between Los Angeles and New York), our species would abandon human space exploration altogether, our leaders contenting themselves with the ridiculous space shuttle, which ventures about as far from Earth as Baltimore is from New York? The feeble broadcasts of the unmanned space probes—the oldest recently passed beyond the edge of the solar system—stir the imagination. But for the most part people have been content to sit at home and wait for the cosmic news to arrive in the form of light from other stars.

We are celestial couch potatoes. Yet what we lack in exploratory will we make up for in other ways. We don't travel with our bodies. We travel with our minds.

Having never left home, the astronomers can say with some confidence that our own galaxy, the Milky Way, is more than 100,000 light-years from end to end and that Andromeda, the galaxy across the street, is 2 million light-years away. These and several other galaxies form our "Local Group." The neighborhood. Just across town are other conglomerations of galaxies like the Sculptor group and the Maffei group, nearly 10 million light-years in distance. A little farther are the Virgo and Fornax clusters, lying some 50 million light-years from the Milky Way. Even if these were miles, the numbers would be staggering. A single light-year is almost 6 trillion miles.

We still haven't left our hometown "supercluster," a galaxy of galaxies a full 200 million light-years across. Beyond it lie more superclusters, stretching to the edge of the visible universe, 10 billion or so light-years from home.

Faced with so grandiose a vision, it is a little surprising to learn that as recently as the 1920s many astronomers thought the Milky Way *was* the universe. Whether there was anything beyond it was a matter of scholarly debate. What are now taken to be vast galaxies similar to our own were dismissed as small nearby gas clouds, insignificant smudges of light.

We were like the villagers in the canyon. Then we discovered a new way to measure.

CHAPTER 1

Black Stars, White Nights

We work from morn till night,
Computing is our duty,
We're faithful and polite,
And our record book's a beauty.
—*From* The Observatory Pinafore

It is only with great difficulty that one can imagine what it was like to be a computer at Harvard Observatory a hundred years ago, not a soulless machine of wire and silicon but a living, breathing young woman. Her name was Henrietta Swan Leavitt and her job was counting stars.

Today this kind of work is done by machine. Arrays of electronic sensors grab images of the sky, long streams of digits for computers to analyze. In the late 1880s, when Harvard embarked on a marathon project to catalog the position, brightness, and color of every star in the sky, the closest things to a modern digital computer were clunky mechanical calculators like the Felt & Tarrant Comptometer or the Burroughs Arithometer, with their rows of clacking buttons, stiff hand-pulled levers, and ringing bells. And there was the human brain. Diligent souls like Miss Leavitt—they actually were called computers—were paid 25 cents an hour (10 cents more than a cotton mill worker) to examine blizzards of tiny dots,

Observatory Hill, 1851 (Harvard College Observatory)

photographs of the night sky. They would measure and calculate, recording their observations in a ledger book.

Imagine a sky with the colors reversed, cold black stars sprayed against a firmament of white. These photographic negatives were produced when a telescope was trained at the heavens, its light focused onto a large glass plate coated on one side with light-sensitive emulsion—a forerunner to photographic film. Today, half a million of these fragile plates are stored in a brick building adjacent to the one where Miss Leavitt and the other computers worked. Fearing an earthquake might shatter this database of glass—the astronomical equivalent of the burning of the library of Alexandria—Harvard built the repository as two nested structures. Physically isolated from the building's exterior shell, an internal matrix of steel beams and flooring rests, the story goes, on an apparatus of leaf springs, like those in a wagon or an old pickup truck.

The result is an invaluable archive of how the sky looked on different nights since the first surveys were done in the 1880s. Among the most precious items in the collection are pictures of the Magellanic Clouds. We know them now as neighboring galaxies, companions to our Milky Way. Back then no one was quite sure what they were. Hunched over the plates in an observatory workroom, Miss Leavitt found the pattern that eventually led to the answer. She discovered a way to measure beyond the galaxy and begin mapping the universe.

Today nearly every scientifically literate person knows, or thinks he knows, that our planet circles an unremarkable star lost among galaxies of galaxies extending billions of light-years in every direction. One can almost hear Carl Sagan intoning the words on public TV. We've learned to revel in our insignificance. As far as most astronomers are concerned, only small details are still in dispute: Is the universe 13.9 billion light-years in radius or just 13.8 billion? So much confidence exudes from these discussions that a spectator may forget to ponder the most basic question: How can anyone know for sure?

Suppose two stars seemingly of equal brightness are shining side by side against the dark dome of night. Knowing nothing else about them, one might conclude that they are equally far away. But that would be true only if the stars happened to be emitting, from their nuclear furnaces, the same amount of light. More likely one star is more powerful than the other yet shining from farther away. How much more powerful and how much farther? Barring a breakthrough in interstellar space travel, there seemed to be no way to find out.

The same uncertainty applied to the faint hazes called nebulae, clouds of light. Were they sprawling galaxies, "island universes" shrunken by their great distance? Or were they little gas clouds right here in the Milky Way? With no means of measuring the universe, the question was almost theological. How many angels can dance on a pinhead? How far away are the stars in the sky?

TODAY A VISITOR to Observatory Hill, a low rise on Garden Street about a fifteen-minute walk northwest of Harvard Square, looks in vain for a sign that anything cosmic happened there. Dwarfed by the giant mountaintop observatories at Palomar, Wilson, Cerro Tololo, and Mauna Kea and blinded by the light of the Boston glare, the Harvard telescope, called the Great Refractor, is now in retirement. But when it saw first light in 1847 it was one of the most powerful in the world.

It had arrived, some liked to say, on the tail of a comet—the March Comet of 1843, which burned so bright that it was visible in broad daylight, a signal to some that the Day of Reckoning was at hand. (A group called the Millerites had used biblical passages to predict that Jesus would make his Second Coming sometime between March 21, 1843, and March 21, 1844. The comet was right on schedule.) Those with a scientific bent felt a deeper kind of awe. Where did the comet come from and when might it return? For answers they could turn to the observatories at Cincinnati or at Yale or Williams colleges. But Harvard didn't have a good enough telescope to properly study the phenomenon. Even the Philadelphia High School was better equipped.

It was an embarrassment Bostonians vowed to correct. Twelve acres called Summer House Hill had recently been purchased by Harvard as an eventual site for a large telescope, but

The Great Refractor (Harvard College Observatory)

little progress had been made. Now the project began in earnest. Well-to-do citizens took out "subscriptions," some $25,000 worth, to build the best observatory in the world. To ensure as stable a platform as possible, the building was constructed around a massive granite pier rooted 26 feet into the ground and rising from the bedrock to the observatory floor. Beneath a 30-foot dome was placed the Great Refractor. The mahogany-veneer tube was outfitted with a lens, 15 inches across, that had been ground by master craftsmen Merz and Mahler of Munich, who were told to make it at least as powerful as the one the Russians had recently purchased for the Imperial Observatory. The space race had begun.

The first astronomer to peer through the glass was rendered nearly speechless: "It is delightful," he wrote, "to see the stars brought out which have been hid in mysterious light from the human eye, since the creation. There is grandeur, an almost overpowering sublimity in the scene that no language can fully express."

With this new tool, astronomers quickly discovered the inner ring of Saturn and, outfitting the telescope with a photographic plate, took the first picture of a star.

2

On a clear night high in the mountains when the air is cold and dry, the brightest stars shine some two hundred and fifty times brighter than the faintest ones, those that can barely be discerned with the naked eye. The ancient Greeks divided this stellar multitude into six categories. The brightest lights were said to be of the first magnitude while the dimmest were of the sixth.

This rough gauge has been refined over the centuries so that each step now means an increase in brightness of about two and a half times. The actual figure is closer to 2.512, conveniently making a fifth magnitude star 2.512 × 2.512 × 2.512 × 2.512 × 2.512 or 100 times dimmer than a star of zero magnitude. A sixth magnitude star is about 250 times dimmer than that, and a seventh magnitude star about 600 times dimmer. (In the original, rather roughshod system, all the brightest stars had been bunched into magnitude one. Measured more precisely, some of them have ended up with magnitudes of less than one, and the brightest with magnitudes of less than 0. Blazing Sirius is about −1.4.)

Centuries ago, with his simple spyglass, Galileo had amplified his vision enough to see stars as faint as the eighth magnitude. The Great Refractor extended the reach to the fourteenth magnitude, resolving images some 400,000 times dimmer than what could be seen from Earth with 20-20 eyes.

With the ability to see farther than ever before, Harvard embarked, in the late 1870s, on the kind of exhaustive search that would become its hallmark, precisely cataloging the brightness of every star in the sky. The observatory was now being run by a young physicist named Edward Charles Pickering, who had made his mark as a professor at the Massachusetts Institute of Technology by establishing the first curriculum in the country where students could confront the ideas of physics head-on—in laboratory experiments, poking at nature and carefully recording the results. He got an early taste of astronomy when he served on two government expeditions to observe total eclipses of the sun. When he was hired in 1876 to take over the observatory, he was thirty years old.

Until this time astronomy had focused on trying to establish two primary details about every star: its position and its motion through space. Pickering was struck by how very little good data had been gathered on two equally important characteristics: a star's precise brightness, a clue to its distance, and color, a clue to what chemicals it contained. Pickering was a fastidious measurer. He occupied himself on hikes in New Hampshire's White Mountains by measuring out the terrain, using an instrument he had fabricated himself. His mission, he decided, and that of the observatory, would be to amass mountains of data, about which others could theorize.

Good old-fashioned astronomy is what he wanted. No big bangs, no black holes, no dark matter—this was way before all that. Space was still flat and of no more than three dimensions. Understanding the universe meant charting little lights as they moved across the sky.

He started with stellar brightness. In the past, astronomers had made some progress along these lines with a German instrument, the Zöllner astrophotometer, which compared a star's brightness to the glow of a kerosene lamp. Focused through a pinhole and reflected by a mirror into the visual field of a telescope, the dot of lamplight appeared as a tiny sun hovering beside a star. The observer adjusted the instrument, stepping down the brightness of the artificial star until, in his judgment, it matched its companion. Then its magnitude could be recorded. (Some of these demanding measurements had been carried out by an observatory employee named Charles Sanders Peirce, who came to be known as one of the most brilliant and eccentric philosophers of all time.)

Pickering felt that a definitive survey should rely on a standard more universal than the brightness of lamplight. He devised an instrument with an arrangement of lenses and mirrors that would allow any star within view to be lined up side by side for comparison with the North Star, which was set, somewhat arbitrarily, at magnitude 2.1. Once astronomers learned how to use the new device, they could knock off as many as one star a minute. Eventually Harvard measured and cataloged forty-five thousand of them.

That was barely a beginning. Within the gaps between stars there were surely many more, so dim and far that they did not register on the retina of an eye, even one fitted with such powerful lenses. To see farther, light from these faint sources would have to be gathered during a time exposure, accumulated on a photographic plate attached to the end of a telescope. Mounted on a rotating platform and driven by mechanical clockworks, the telescope could track a star as it arced across the sky, pooling its light photon by photon, chemically etching an impression.

The astronomical leverage this provided was stunning. From Earth the Pleiades appear as a subtle glow engulfing seven bright points of light—the "Seven Sisters" of Greek mythology, pursued by Orion. Galileo had already seen through his telescope that the sisters were joined by dozens more. A three-hour time exposure, taken in Paris, revealed that the cluster included more than 1,400 stars.

More stars still could be unveiled by mounting telescope and camera as far above sea level as possible, cutting through miles of atmospheric distortion. After unsuccessful attempts to establish stations at Pikes Peak in the Colorado Rockies and Mount Wilson in Southern California, Pickering decided to try the high reaches of Peru. He dispatched an expedition led by a trusted colleague, Solon I. Bailey, who established a temporary post atop a peak that, it was decided, would now be called Mount Harvard. Bailey hadn't reckoned on the length of the annual rainy season and was forced by the clouds to find a clearer site, finally settling in a remote town called Arequipa. This time the location seemed perfect. Pickering arranged to have an observing station sent piece by piece, from Boston Harbor around the tip of South America. Included among the cargo were the components of the 24-inch Bruce Telescope (named for the heiress who paid for its construction, Catherine Wolfe Bruce).

For all Pickering's hopes, the project got off to a bad start. His brother William, as headstrong and arrogant as Edward was modest and reserved, was placed in charge of Arequipa, mismanaging the operation and scandalizing the world astronomical community when he began dispatching outlandish scientific reports to that great academic journal the *New York Herald*. Ignoring his assignment to study the stars, he trained the telescope on Mars, enthusiastically describing huge mountain ranges rising above giant rivers and lakes extending hundreds of square miles—a geography that remained stubbornly invisible to any eyes but his own.

While Edward Pickering concentrated on damage control at home, Bailey was sent back to Peru to retake the observatory. Before long, the Arequipa station was shipping crate after crate of photographic plates north to Cambridge, the first pieces of what would become a mosaic of the entire southern sky.

With so much new information to digest, astronomers were soon overwhelmed. They were faced with an embarrassment of riches now familiar throughout science, a burgeoning glut of undigested data begging to be categorized. That is where the computers came in.

3

"A great observatory should be as carefully organized and administered as a railroad," Pickering once observed. "Every expenditure should be watched, every real improvement introduced, advice from experts welcomed, and if good, followed, and every care taken to secure the greatest possible output for every dollar expenditure. A great savings may be effectuated by employing unskilled and therefore inexpensive labor, of course under careful supervision."

Edward Pickering (Harvard University Portrait Collection)

Imagine trying to find people to do such precise work for 25 cents an hour—what amounted to the minimum wage. Today the job would probably have to be farmed out to star-counting sweatshops in Asia. For the late nineteenth century, computing wasn't such a bad deal. Seven hours a day, six days a week, the job paid $10.50 a week and included a month's vacation. Not many men were interested in the tedious work, so the positions went mostly to women. (The tradition was a long time in passing. As recently as the early 1960s, Brookhaven National Laboratory hired Long Island housewives to pore over the tangled images of subatomic particles, looking for patterns that might foretell a new physics.)

Recognizing that his housekeeper, Williamina Paton Fleming, was overqualified for mopping floors, Pickering hired her as one of his first computers. (Abandoned by her husband after emigrating from Scotland, she was grateful enough to name her son, born that year, Edward Pickering Fleming.) She eventually became curator of the collection of photographic plates, doubling her salary, and was in charge of classifying stars according to their spectra, the colors revealed when their light was refracted through a prism. This was another of the observatory's ambitious efforts, resulting in a monumental work called the Henry Draper Catalogue, named for an accomplished and wealthy amateur astronomer who had taken the first photograph of a nebula. Funded by Draper's widow, the compendium provided employment for two other computers, Annie Jump Cannon and Antonia Caetana Maury.

This women's work sometimes resembled bookkeeping more than scientific research. Pickering tried to make it reasonably stimulating and treated his computers with respect. But he was intent on the observatory's getting its money's worth.

"He seems to think that no work is too much or too hard for me no matter what the responsibility or how long the hours," Mrs. Fleming complained in a diary. "But let me raise the question of salary and I am immediately told that I receive an excellent salary as women's salaries stand."

> If he would only take some step to find out how much he is mistaken in regard to this he would learn a few facts that would open his eyes and set him thinking. Sometimes I feel tempted to give up and let him try some one else, or some of the men to do my work, in order to have him find out what he is getting for $1,500 a year from me, compared with $2,500 from some of the other assistants.

> Does he ever think that I have a home to keep and a family to take care of as well as the men? But I suppose a woman has no claim to such comforts. And this is considered an enlightened age! . . . I feel almost on the verge of breaking down.

When she asked him for a raise, Pickering agreed to pass on the request to the university president. Money was always tight. This was before the era of government-funded big science, and observatories were dependent on the charity of rich benefactors, and on people with a monastic dedication to the craft. Pickering worked as hard as any of them, administering by day, stargazing by night. When the sky was cloudy he would do calculations late into the evening, sometimes with an assistant reading to him for entertainment (Shakespeare was a favorite). Considering the long hours, his $3,400 annual salary came to much less than two dollars an hour. (He and his family also got to live in the less-than-luxurious director's residence on Observatory Hill.) No one was in this for the money.

"An astronomer is a sorry soul," began a chorus in *The Observatory Pinafore*, a parody of Gilbert and Sullivan's comic operetta *H.M.S. Pinafore* written by one of Pickering's assistants.

> *He must open the dome and turn the wheel,*
> *And watch the stars with untiring zeal,*
> *He must toil at night though cold it be,*
> *And he never should expect a decent salaree.*

Most of the time, the computers seemed to enjoy their jobs, making light of the low pay and somewhat Dickensian working conditions. In *The Observatory Pinafore*, one of them,

The Observatory Pinafore (Harvard College Observatory)

"Josephine," sings of her toil in the "dark and dingy place, all cluttered up and smelling strong of oil," apparently from the furnace that had recently replaced fireplaces for taking the edge off the New England cold. At another point in the story, a whole chorus of computers breaks out in song:

We work from morn till night,
Computing is our duty,
We're faithful and polite,
And our record book's a beauty.

It is tempting to imagine Henrietta Swan Leavitt joining in the song. But that couldn't be. Though written in 1879 the musical was not actually performed until New Year's Eve 1929. By then she had already died.

CHAPTER 2

Hunting for Variables

My friends say, and I recognize the truth of it, that my hearing is not nearly as good when absorbed in astronomical work.

—Henrietta Leavitt in a letter to Edward Pickering

Although unmarked by a plaque, the second-floor room where Miss Leavitt and the other computers probably worked is still intact. The university, always pressed for space, hasn't been as diligent about preserving the old observatory buildings as it has the collection of photographic plates. The wooden ceiling beams have been painted over in institutional white enamel. Fluorescent lights have been retrofitted where chandeliers once hung and an air conditioner mounted in one of the old sash windows. The place has all the charm of a room in a state hospital. Nearby a dumbwaiter, which shuttled the glass plates up from below, now holds a computer printer. A closet is crammed full of abandoned IBM Selectric typewriters, another archeological layer of cast-off technologies.

Haul off the junk, restore the late-nineteenth-century decor, and imagine Miss Leavitt, as she would have expected to be called, in a long frilled dress buttoned to the neck, her dark hair pulled tightly into a bun (we are extrapolating here from one of the only existing photographs). She is sitting at a table

Henrietta Swan Leavitt (Harvard College Observatory)

before a wooden viewing frame that supports a large glass plate—one of those black-on-white reversals of the night sky. At the base of the frame is a mirror, reflecting light in from a nearby window to illuminate the image from behind. Around her sit other computers, similarly occupied, and occasionally Edward Pickering drops by to see how the calculations are going.

She was twenty-five when she arrived at the observatory in 1893 as a volunteer. Her goal was to learn astronomy, and apparently she was of somewhat independent means. The daughter of a Congregationalist minister, George Roswell Leavitt, Henrietta had been born on the Fourth of July 1868 in Lancaster, Massachusetts, to what once was called "good Puritan stock." Her ancestry can be traced back to Josiah Leavitt of Plymouth, and to four centuries of Leavitts in Yorkshire, England.

At the time of the 1880 census the family was living in one half of a large double house, 9 Warland Street in Cambridge, near Pilgrim Congregational Church, at the corner of Magazine and Cottage streets. Reverend Leavitt served as pastor. The neighborhood was solidly middle to upper-middle class. The Leavitts' neighbors included a piano tuner, a clerk, a police captain, a grammar school teacher, and a civil engineer, as well as a soda and mineral water manufacturer, a carriage manufacturer, and the owner of a plumbing company.

When he dropped in to take the census, the enumerator found Mrs. Leavitt, also named Henrietta Swan (her maiden name was Kendrick), tending to three of her children, George, Caroline, and two-year-old Mira. (It was the last year of her life. A few months later Mira was dead.) Eleven-year-old Henrietta, the eldest, and another sister, Martha, were away at school. Another brother, Roswell, had died in 1873, when he was fifteen months old; the youngest, Darwin, would be born two years later.

The household must have felt crowded. Henrietta's aunt, Mary Kendrick, also lived there, as did a servant girl. Next door, in No. 11, her grandfather, Erasmus Darwin Leavitt, lived with his wife and a thirty-year-old daughter (the census taker noted that she had a spinal injury). They too employed a live-in servant.

The family valued scholarship. Henrietta's father had graduated from Williams College and earned a doctorate in divinity from the Andover Theological Seminary. Like her grandfather, an uncle (Reverend Leavitt's brother) was named Erasmus Darwin, perhaps after the renowned eighteenth-century physician and naturalist, the grandfather of Charles Darwin. The younger Erasmus, the second president of the American Society of Mechanical Engineers, would gain national prominence for his design of the Leavitt pumping engine at the Boston Water Works' Chestnut Hill station. He was also a fellow of the American Academy of Arts and Sciences.

A few years later the Leavitts moved to Cleveland, and in 1885 Henrietta enrolled at Oberlin College, taking a preparatory course followed by two years of undergraduate work. Returning to Cambridge in 1888, she entered Radcliffe, then called the Society for the Collegiate Instruction of Women. (One of her cousins, a daughter of Erasmus Leavitt, was in the same freshman class.)

The entrance requirements were strict. Every young woman was expected to prove her familiarity with a list of classics—Shakespeare's *Julius Caesar* and *As You Like It*; Samuel Johnson's *Lives of the Poets*, William Makepeace Thackeray's *English Humorists*; Swift's *Gulliver's Travels*; Thomas Gray's poem "Elegy Written in a Country Churchyard"; Jane Austen's *Pride and Prejudice*; and Sir Walter Scott's *Rob Roy* and "Marmion"—by writing, on the spot, a short composition. There were also tests on language ("Translation on sight of simple [in the case of Latin, Greek, and German] or ordinary [in the case of French] prose"), on history ("Either [1] History of Greece and Rome; or [2] History of the U.S. and of England"), mathematics (algebra through quadratic equations and plane geometry), and physics and astronomy. In addition to these examinations on "elementary studies," students had to demonstrate more advanced knowledge in two subjects—mathematics, for example, and Greek. The catalog noted reassuringly, "A candidate may be admitted in spite of deficiencies in some of these studies; but such deficiencies must be made up during her course."

The only deficiency listed on Henrietta's transcript was in history, which she had corrected by her junior year. Along the way she also took courses in Latin, Greek, the humanities, English, and modern European languages—German (her only C), French, and Italian—and in fine arts and philosophy. She didn't take, nor was offered, much science: just natural history, an introductory physics class (she got a B) and a course in analytic geometry and differential calculus (an A). It was only in her fourth year that she enrolled in astronomy, earning an A–. Observatory Hill is just up Garden Street from Radcliffe, and some of the Harvard astronomers, supervised by Edward Pickering, taught classes there.

In 1892, shortly before her twenty-fourth birthday, Henrietta graduated with a certificate stating that she had completed a curriculum equivalent to what, had she been a man, would have earned her a bachelor of arts degree from Harvard. She stayed in Cambridge and the next year was spending her days at the observatory, earning graduate credits and working for free. Maybe her uncle Erasmus had pulled some strings for her. She stayed there two years.

No diary has been found recording what it was about the

stars that moved her. One of history's small players, her story has been allowed to slip through the cracks. She never married and died young, and it is only upon her death that we find, in an obituary written by her senior colleague Solon Bailey, testimony to what she might have been like as a woman:

> Miss Leavitt inherited, in a somewhat chastened form, the stern virtues of her puritan ancestors. She took life seriously. Her sense of duty, justice and loyalty was strong. For light amusements she appeared to care little. She was a devoted member of her intimate family circle, unselfishly considerate in her friendships, steadfastly loyal to her principles, and deeply conscientious and sincere in her attachment to her religion and church. She had the happy faculty of appreciating all that was worthy and lovable in others, and was possessed of a nature so full of sunshine that, to her, all of life became beautiful and full of meaning.

Although the obituary didn't say so, she was also deaf, although apparently not from birth. In her second year at Oberlin she had enrolled as a student in its conservatory of music. For her new calling, eyes were more important than ears, and perhaps deafness was an occupational advantage in a job requiring such intense powers of concentration.

2

Pickering, always happy to have a motivated volunteer, put her to work recording the magnitude of stars, a craft called stellar photometry. During a time exposure, brighter stars leave larger spots on a photographic plate, chemically darkening more grains in the emulsion. Size therefore is an indicator of brightness. Looking through an eyepiece, Miss Leavitt would compare each pinpoint against stars whose magnitudes were already known. Sometimes this information was displayed on a small palette marked with spots arranged and labeled according to magnitude. Shaped like a miniature flyswatter, the tool was called a "fly spanker." When she was satisfied that she had gauged the brightness correctly, she would record the information in neat tiny writing on a sequentially numbered pink-and-blue-striped page in a ledger book initialed "HSL."

Early on she was asked to look for "variables," stars that waxed and waned in brightness like slow-motion beacons. (A few of the more interesting were to be found in a constellation she might have considered her namesake, Cygnus, the Swan.) Some of the variables completed a cycle every few days, others took weeks or months. Her job was not to speculate why. For a while it was thought that each variable actually consisted of two stars orbiting about a common point, periodically eclipsing each other's light. More recent evidence indicated that the temperature (judged by the color of the starlight) rose and fell along with the brightness. That suggested that these variables were probably single stars that periodically flared and dimmed. The reasons for the pulsation remained obscure. This was before anyone knew what powered the stars, much less why the flame might not burn steadily.

In any case, the rhythms were imperceptibly slow and subtle. (Astronomers were surprised to learn later that Polaris, the North Star, which they had been using as their touchstone for measuring brightness, regularly varied in magnitude.) Only by measuring stars at various intervals through the year could one detect the variations.

Photography made that possible. With dense swarms of stars on every plate, it was impossible to check each one. To find likely prospects, a researcher would take two plates of the same region exposed at different times and line them up, sandwich style, one on top of the other. One image was the standard black-star negative produced by the camera. The other was a positive. Align them just so and they would cancel each other—except for the stars that had changed in brightness. They would look subtly different. A black dot surrounded by a white halo would mean, for example, that a star had brightened, its image on the plate expanding in size. If a star seemed particularly promising, more plates would be compared. (If necessary Pickering would order new ones from the astronomers.) Plate by plate, the computers would measure a dot as it swelled and receded, writing tiny numbers on the glass in india ink.

Henrietta Leavitt spent day after day doing this painstaking work, absorbing herself in the data with what one colleague later called "an almost religious zeal." She wrote up a draft of her findings, then sailed for Europe in 1896, where she traveled for two years. We don't know where or with whom.

She hadn't forgotten about astronomy. When she landed back in Boston, she conferred briefly with Pickering, who suggested some revisions to her work. Then, taking the manuscript with her, she left for Beloit, Wisconsin, where her father was now the minister of another church. Because of personal problems, never really explained, she remained there more than two years, working as an art assistant at Beloit College. She must have found the work unsatisfying, for on May 13, 1902, she wrote to Pickering, apologizing for letting her research languish and for being out of touch for so long. She hoped he would let her resume her projects from Wisconsin.

"The winter after my return," she explained, "was occupied with unexpected cares. When, at last, I had leisure to take up the work, my eyes troubled me so seriously as to prevent my using them so closely." Her eyes were strong now, she assured him, and her interest in astronomy had not waned. She asked whether he might send her the notebooks she needed to complete her manuscript. "I am more sorry than I can tell you that the work I undertook with such delight, and carried to a certain point, with such keen pleasure, should be left uncompleted. I apologize most sincerely for not writing concerning the matter long ago."

She also mentioned having some trouble with her hearing, worrying, a little oddly, that stargazing might make it worse. "My friends say, and I recognize the truth of it, that my hearing is not nearly as good when absorbed in astronomical work." Cold weather seemed to aggravate her condition. "It is evident that I cannot teach astronomy in any school or college where I should have to be out with classes on cold winter nights. My aurist forbids any such exposure.

"Do you think it likely that I could find employment either in an observatory or in a school where there is a mild winter climate? Is there anyone beside yourself to whom I might apply?"

It seems from the letter that her hearing problems then were fairly mild—the first ailment she mentions is with her eyes. Yet one of the standard reference books, *Dictionary of Scientific Biography*, states that as early as Radcliffe she was extremely deaf—a misapprehension that has been picked up and recycled again and again.

She must have been delighted with Pickering's response. Three days later, he offered her a full-time job. "For this I should be willing to pay thirty cents an hour in view of the quality of your work, although our usual price, in such cases, is twenty five cents an hour," he wrote. If it was not possible for her to relocate, he would pay her fare for a short visit to Cambridge. She could get her work in order to take home to Beloit.

"I do not know of any observatory in a warm climate, where you could be employed on similar work," he continued, "and it would be difficult to furnish you with a large amount of work that you could carry on elsewhere." In any case, he noted, "I should doubt if Astronomy had anything to do with the condition of your hearing, unless you have been assured that this is the case by a good aurist."

She gratefully accepted his proposal for a working visit to Observatory Hill. "My dear Prof. Pickering," she replied, a few days later. "It has proved possible for me to arrange my affairs here so that I can go to Cambridge next month and remain until the work is completed. Your very liberal offer of thirty cents an hour will enable me to do this." She planned to arrive around the time of Harvard commencement and to take up her job before the first of July.

En route to the observatory, she stopped in Ohio to visit relatives, encountering another of the family problems that seemed to punctuate her life. "The illness of a relative with whom I stopped to visit on my way to Cambridge is likely to detain me for some time," she wrote to Pickering, adding that she might be waylaid for several weeks. "I am sorry for the delay, which seems to be unavoidable, but my face is set toward Cambridge and I hope it may not be very long before I can report at the Observatory."

Finally, on August 25, she contacted him again from an

address in Brookline, Massachusetts, where she was temporarily staying, announcing her arrival.

"It has been a disappointment to me that I have had to defer the beginning of my work for so long," she wrote. "At last I find myself free to take it up, and expect to go to the Observatory Wednesday afternoon, arriving there between half-past two and three. If there is a time which will suit you better, will you please let me know either by letter or by telephone?" At first she could work only about four hours a day, but was hopeful that she would soon be able to give him her full attention. "I hope that this long delay has in no way inconvenienced you." Whether Pickering was beginning to find these plaints a bit wearing has gone unrecorded. In the coming years there would be many more.

Working through the fall semester, she took winter break again in Europe. A letter dated January 3, 1903, finds her aboard the S.S. *Commonwealth*, a mail steamer for the Dominion Line, writing to Williamina Fleming about a misdirected paycheck. (Henrietta's younger brother, George, who was living in Cambridge with Uncle Erasmus, now a widower, helped straighten out the problem.) A year earlier the shipping line had initiated winter Boston-to-Mediterranean service; so perhaps she was off with a friend to Italy, or the South of France. It is nice to imagine her on deck at night, bundled up to keep her aurist happy, looking up at the stars.

CHAPTER 3

Henrietta's Law

What a variable-star "fiend" Miss Leavitt is—One can't keep up with the roll of the new discoveries.

—A colleague in a letter to Edward Pickering

During the first circumnavigation of the Earth, Ferdinand Magellan's crew relied on "certain shining white clouds" to find its way. There is no South Star to navigate by down under but the Nubecula Major and Nubecula Minor—the big and little clouds, as Magellan called them—helped one maintain a steady course. Invisible in most of the Northern Hemisphere, these impressive formations came into view as the fleet reached the latitude of what is now Brazil.

Magnified by telescopes, the pair of luminous hazes dazzled the mind. "In no other portion of the heavens are so many nebulous and stellar masses thronged together in an equally small space," wrote the astronomer John Herschel after observing them from the Cape of Good Hope in 1834. It was as though two pieces of the Milky Way had broken off and drifted. But that begged the question: Were these separate and distant galaxies or something smaller and nearby, suburbs of the Milky Way?

There was no reason to hope that the answer would be

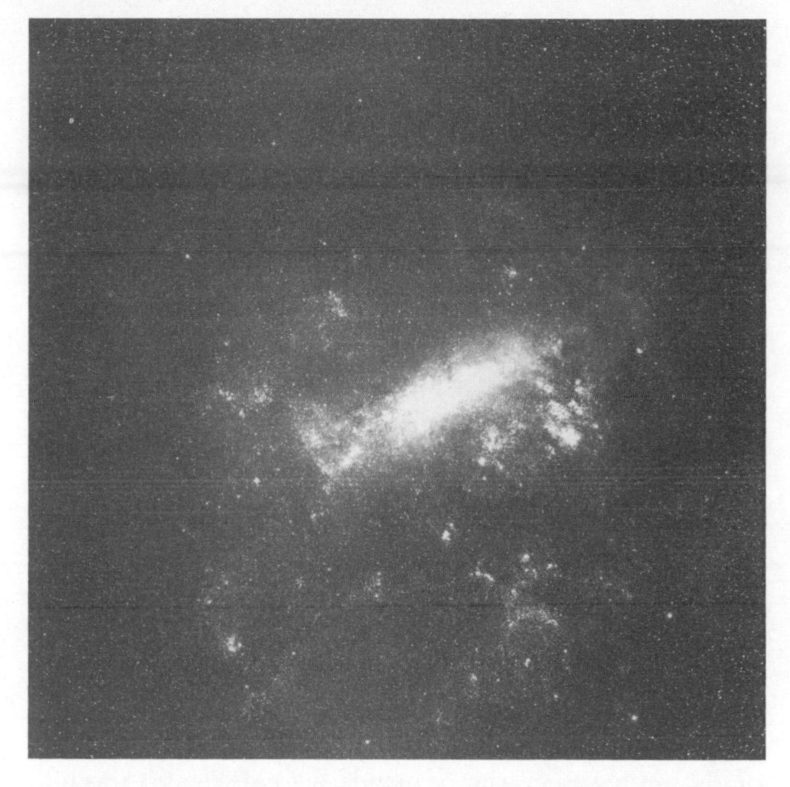

The Large Magellanic Cloud (Harvard College Observatory)

found in the images of the Magellanic Clouds that had been photographed at Harvard's station in Arequipa, Peru. Miss Leavitt was charged with nothing more than examining the plates for variable stars. Each time she spotted one, she would scrutinize it with a magnifying eyepiece, determine its coordinates on the photographic plate, and carefully compute its change in magnitude by comparing it with other stars.

While she worked she might have remembered a story her father once told about Herschel himself. Speaking at an annual meeting of the American Missionary Association, Reverend Leavitt described how astronomers throughout Europe had greeted their colleague's great discoveries with disbelief: "Men said to him, in angry letters, 'We do not see what you see.' " Herschel, Reverend Leavitt continued, was ready with a response: "Perhaps you do not take the care in your observations that I do. . . . [W]hen I observe on a winter night I place my glass on the lawn at Greenwich, and let it stand there until the instrument comes to be of the temperature of the air.' " Moreover, the reverend said, Herschel ensured that his own body temperature would not affect the observations. " 'Oftentimes,' he said, 'I have been out in the winter air for two hours before I would open my glass, because I must come to be of the same temperature as the instrument itself.' "

Reverend Leavitt found a rather oblique spiritual message in the story, something about how a preacher must be of the same spiritual "temperature" as his "instrument" (God's word), and of the Bible and the heavens as well. Herschel's words might serve better as advice to a young astronomer: for all the excitement of discovering a new star or nebula, a good scientist was ultimately one who took care to consider the minutest of details, weighing each tiny component that made up an observation.

Henrietta apparently found such meticulous work satisfying enough that she amended her original plan to take her work back to Wisconsin. After a voyage to the British Isles in the summer of 1903 on the H.M.S. *Ivernia*, she took a quick train trip to Beloit to prepare for her relocation to Cambridge as a permanent member of the observatory staff.

Her decision paid off. One spring day in 1904 she was comparing plates of the Small Magellanic Cloud, taken at different times, when she noticed in the stellar spray several dots that had swelled and then receded in size. Variables. Her interest piqued, she examined other images, finding dozens more.

That fall sixteen more plates of this nebula were served up by the astronomers at Arequipa and shipped north to Boston, arriving at the observatory in January. When Miss Leavitt began scrutinizing these new photographs, variables popped up one after another—"an extraordinary number," she later wrote. The results, published in the observatory's regular circulars, made an immediate impression.

"What a variable-star 'fiend' Miss Leavitt is," a Princeton astronomer wrote to Pickering. "One can't keep up with the roll of the new discoveries." Even the newspapers took notice. A column of flippant news briefs in the *Washington Post* noted, tongue in cheek: "Henrietta S. Leavitt, of the Harvard Observatory, has discovered twenty-five new variable stars. Her record almost equals Frohman's." (Charles Frohman was a powerful theatrical producer and booking agent.)

Day after day, she quantified the specks of pulsing starlight, filling in column after column of numbers. If there was anything noteworthy or unusual about a variable she would add a comment. The star with Harvard Number 1354 was "the northern star of a close pair, in a group of five." Number 1391 was "the southern star in a line of three." Number 1509 "appears to be at the centre of an extremely small, faint cluster."

Each star was an individual. Before long, she had discovered and cataloged hundreds of them in the two Magellanic Clouds, some of which flared no brighter than the fifteenth magnitude—thousands of times dimmer than the faintest stars she might have seen on a particularly clear night in the New England countryside.

She was boarding with Uncle Erasmus in a large Italianate

villa (now part of the Longy School of Music) recently built for him on Garden Street. The house was just a short walk from Observatory Hill, where, for the next few years, she continued her research. Piece by piece, her results appeared in brief progress reports sometimes given to Pickering or Bailey to read in her absence at the December meetings of the Astronomical and Astrophysical Society of America, when she would be home in Beloit for Christmas. By 1908, six years after she had resumed her work, she published a full account, "1777 Variables in the Magellanic Clouds," in the *Annals of the Astronomical Observatory of Harvard College*. Twenty-one pages in length, the paper included two plates and fifteen pages of tables.

The sheer number of variables was surprising enough. But a reader with the patience to make it to the end of the paper would have found something even more remarkable. Almost as an afterthought, she had singled out sixteen of the stars, arranging them in a separate list showing both their periods and their magnitudes. "It is worthy of notice," she observed, that "the brighter variables have the longer periods."

In light of what astronomers know now, this is an understatement as magnificent as the one Watson and Crick made at the end of their famous 1953 paper on the double-helical structure of DNA: "It has not escaped our notice that the specific pairing we have postulated immediately suggests a possible copying method for the genetic material." This when they were telling their friends that they had discovered the secret of life.

Miss Leavitt was not being coy. She just didn't want to overinterpret her data. Since the variables were all in the Magellanic Clouds, they must be roughly the same distance from Earth. If the correlation she glimpsed held true, you could judge a star's true brightness from the rhythm of its beat. Then you could compare that with its apparent brightness and estimate how far it was. This was too profound a conclusion to hang on just sixteen stars. More measurements would have to be made.

2

That wasn't to happen anytime soon. The same year her results were published, Henrietta fell ill. On December 20, she wrote to Pickering from a Boston hospital, where she had been confined for the past week, thanking him for "the beautiful pink roses" and "for the kind thought so beautifully expressed. It means much at a time like this to be made to realize that one is remembered by one's friends."

To convalesce, she returned to Wisconsin to stay with her parents and two unmarried brothers, George, now a missionary, and Darwin, another clergyman. After resting through the next spring and summer, she planned to resume work in the fall. But in September she reported to Pickering that a "slight illness" contracted after a visit to a lake near Beloit had "proved unexpectedly obstinate, and I cannot tell when I shall be able to get away."

In October, after she had been absent for close to a year, Pickering wrote to inquire whether she would like him to send her some work. By early December, when she had not responded, he asked again, this time letting a glint of impatience show. "My dear Miss Leavitt," he began. "It is with much regret that I hear of your continued illness. I hope you will not undertake work here until you can safely do so. It may however relieve your mind if we can dispose of two or three questions. . . ."

First he asked if she would send a letter at the beginning of each month stating whether she would be returning any time soon. Then he proposed that she issue a brief report (a so-called *Harvard College Observatory Circular*) describing the preliminary results of another study she had been engaged in, the North Polar Sequence, a Herculean effort to measure, more accurately than ever before, the magnitudes of ninety-six stars near Polaris. This was one of Pickering's pet projects, of a higher priority to him than her study of variables. He hoped the North Polar Sequence would become the gold standard for gauging the brightness of stars throughout the sky.

She replied three days later, apologizing for being too weak to answer his earlier letter. "I thank you for expressing the desire that I wait for complete recovery before returning to Cambridge; it would make it even harder for me to be idle than it now is, if pressure were brought from without, especially from you. The thought of uncompleted work, particularly of the Standard Magnitudes, is one I have had to avoid as much as possible, as it has had a bad effect nervously." If she was just as anxious about her Magellanic variables, she didn't say.

She held out hope that her condition might improve enough for her to resume work after Christmas. "Not the least of my trial in being ill is the knowledge of the annoyance it causes you."

In mid-January, Pickering wrote again, opening with the now familiar greeting and lament. "My dear Miss Leavitt: It is with much regret that I hear that your illness will again require you to postpone your return to Cambridge. It occurs to me that, when you do return, you may be able to do much of your work in your room, and thus save yourself the walk to the Observatory."

In the meantime she had told Mrs. Fleming that she was ready to work from Beloit, and Pickering, emphasizing again the importance of the North Polar Sequence, described some photographs he planned to send her, including one from the Mount Wilson Observatory in California, where a new 60-inch telescope, the largest in existence, was recording stars of exceeding faintness. He outlined his thoughts on how she should proceed with her measurements. "What do you think of this plan and can you suggest any improvements in it? . . . I hope you will not let these matters trouble you, and that you will not undertake any of this work, except with the approval of your doctor."

A few weeks later she received a box from Cambridge packed with photographic plates, paper prints, ledgers, a wooden viewing frame, and a 1½-inch eyepiece, allowing her to get back to gauging magnitudes. She responded with assurances that she was "now strong enough to work for two or three hours a day, and am very glad indeed to have the means of employing the time to advantage." She hoped, as always, for an early return to Cambridge.

For the next three months, she continued her calculations, sending back detailed reports to Observatory Hill. Her health continued to improve but at a glacial pace both she and Pickering must have found exasperating. "It is a great pity that my latest attempt to fix a time for returning to Cambridge should have failed like the others," she wrote to him in April, almost a year and a half since her absence began. This time, she assured him, her return really was imminent. "My physician has not yet given his consent to my departure, wishing to be assured of the soundness of my recovery. I now expect to receive my dismissal from him any day and to be in a position to make definite plans for resuming work."

By May 14, 1910, she was finally back in Cambridge, or was at least on her way. Her name and that of her colleague Annie Cannon appear on a list of observatory employees requesting tickets to the annual Harvard Class Day commencement ceremonies.

Her homecoming was short-lived. The following March her research was interrupted again when her father died, leaving his widow a modest estate that after probate costs and the settlement of debts was valued at just over $9,000. (He had kept the house on Warland Street, where Henrietta had been a girl, and owned a small amount of stock in a copper mining company that his brother Erasmus consulted for.) After thanking her colleagues for sending flowers, Henrietta departed for Beloit to console her mother.

When she had not returned by June, Pickering sent her a box with seventy photographic plates and other material for the North Polar Sequence, but she wasn't able to concentrate on the work for long. Ten days later she wrote to inform him that she and her mother were departing "rather unexpectedly" to stay with some in-laws in Des Moines. She offered no explanation.

She took the plates to the Beloit College library for safekeeping. "It is a new, fire-proof building," she assured him, "and is to be open all summer. The plates are on a strong shelf in a corner of the librarian's private office, and are labeled with a request that no one shall touch them. Orders to that effect have been given to the janitor. . . . It will be a disappointment to lose nearly a month in my work on the plates, but there will be a good deal of work with the papers I shall take with me."

Her research wasn't entirely neglected. She even found time for her variable stars, sending a report for Pickering to read at a conference in Ottawa. After several more delays and apologetic letters, she returned that fall to her uncle's house on Garden Street.

Finally given the luxury of a long stretch of uninterrupted time, she turned back to the strange matter of the Magellanic variables, plotting twenty-five of them on a graph with their brightness on one axis and their period on the other. Her results were published in 1912 in a *Harvard Circular* under the name of Edward Pickering: "The following statement regarding the periods of 25 variable stars in the Small Magellanic Cloud has been prepared by Miss Leavitt."

The pattern now seemed clearer than ever. The stars lined up so neatly that she was moved almost to exclamation: "A remarkable relation between the brightness of these variables and the length of their periods will be noticed." The brighter the star, the slower it blinked. Why she didn't know, and for now it didn't matter. "Since the variables are probably at nearly the same distance from the Earth their periods are apparently associated with their actual emission of light."

In other words, you could determine how bright they really were. Without leaving Earth, you could count the beats of a star's rhythm, then use this to calculate its intrinsic magnitude. Compare that with its apparent magnitude and you would have its distance.

The universe had provided, for the especially keen observer, a hint of its grandeur. Imagine that you are standing on a back porch at night looking out over a dark field. Somewhere on the far edge is a mysterious array of electric lights. Some are brighter, some fainter, but since you don't know how bright they really are, you can't tell whether they are ten yards or ten miles away.

Now suppose that the lights are blinking, and that it has been decreed by some international authority that bulbs be manufactured so that they flash according to their brightness.

Fifty-watt bulbs blink faster than 100-watt bulbs. If two of the beacons are pulsing at the same frequency, you know they are equally bright. So if one appears, say, four times dimmer, it has to be farther away.

To be precise, it is two times farther. Light traveling through space spreads and diminishes according to the inverse square law. Square the difference in distance and you get the difference in brightness. All other things being equal (which they never quite are), a light that is nine times dimmer than another must be three times farther away.

Although Miss Leavitt didn't use the term in her paper, the variables with this remarkable property are called Cepheids, for the first was discovered, in 1784, in the constellation Cepheus by an amateur English astronomer named John Goodricke. (He and Henrietta had more than astronomy in common: Goodricke was deaf and she was steadily becoming so.) The new law linking period and brightness would become known throughout astronomy as the Cepheid yardstick, a way to measure through great stretches of space.

There was just one problem: her Cepheids revealed only relative distances. You could say with some confidence that one star was twice as far away as another, and three times farther than another. But were they one, two, and three light-years from earth, or twenty, forty, and sixty? There was no way to know. To turn the ratios into actual distances, someone needed to discover how far the closest of the stars is from Earth.

For now, Miss Leavitt's new yardstick was one without numbers. The next step would be to calibrate it.

CHAPTER 4

Triangles

I had not thought of making the very pretty use you make of Miss Leavitt's discovery about the relation between period and absolute brightness.

—Henry Norris Russell, in a letter to Ejnar Hertzsprung

Hold a finger a couple inches from your nose and alternately blink from left eye to right. The finger rapidly changes position, and the farther you move it from your face the smaller the displacement becomes. The window frame on the wall shifts even less and the telephone pole down the block hardly at all. The separation between the eyes—astronomers and surveyors call it the baseline—is too small. We can sense the depth of the world close by, but the farther we look the flatter it appears. The distant mountains might as well be cut from cardboard.

Wired by nature to register this effect (remember it's called parallax), the brain carries out a kind of neurological trigonometry. The two eyes and the object before them form a triangle, and you unconsciously compute the distance from base to apex. Without even thinking about it, you triangulate.

Get up now and walk from one side of the window frame to the other, a few feet away. The size of your baseline has increased, and now the telephone pole seems to shift against the building standing behind it. The effect is the same as having a very wide head, an eye at each side of the window looking out from two different angles. Measure the separation between the two observation points—the base of the imaginary triangle and the angles formed by each window and the pole. You can do this with a surveyor's transit (think of it as a kind of protractor). With just this information and some high school trigonometry, you can calculate the altitude of the triangle—the distance to the telephone pole. The calculation itself needn't concern us. Just know that the very nature of triangles ensures that they can be completely defined with only three scraps of information—two angles and one side or one angle and two sides. All the other dimensions follow from that.

With a wider baseline—an eye on each side of the block—you could measure how far the building is. With a long enough stretch, even the mountains on the horizon will seem to move.

The history of astronomical measurement preceding Henrietta Leavitt's discovery can be compressed into a story of how people learned to use larger and larger triangles to point farther into the sky.

SEND TWO OBSERVERS to different places on the Earth's surface and, if the separation is great enough, each will see the moon at a slightly different position against the backdrop of stars. Have them measure both angles simultaneously (they would need to synchronize their watches) and, if you know the length of the baseline, you can triangulate.

In a variation, the Greek astronomer Hipparchus in the second century B.C. used a solar eclipse as his timepiece. While the sun was blotted out completely at Hellespont, the strait in northwestern Turkey near the ancient city of Troy, the eclipse

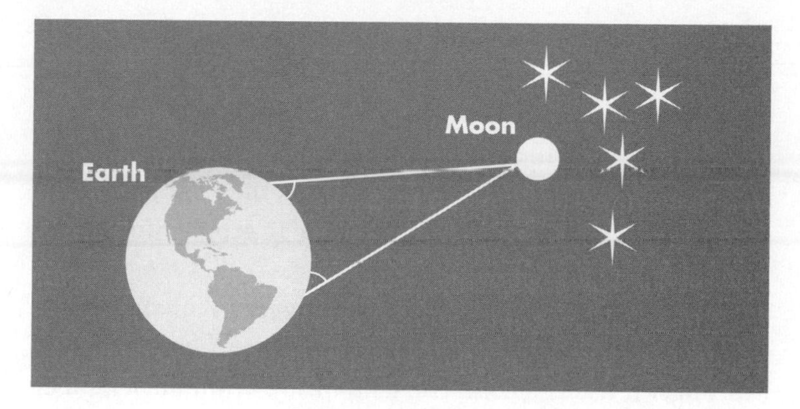

Lunar Parallax

was only four-fifths full in Alexandria. If you could freeze time and hop back and forth between these two vantage points, the moon would seem to shift by an amount equal to one-fifth the size of the solar disk—the Earth blinking its eyes. The sun occupies about half a degree of the arcing bowl of sky, making the parallax of the moon one-tenth of a degree.

Had Hipparchus known the distance between Troy and Alexandria—the size of the baseline—he would have had his answer. Instead, using a more complex arrangement of triangles, he was able to calculate that the distance of the moon must be about thirty times greater than the diameter of the Earth. He got the ratio right. The planet has since been measured at about 8,000 miles across. Using that number, Hipparchus's method would put the moon 240,000 miles away. Two thousand years later, people learned to make the measurement by bouncing radar signals off the moon and measuring the delay of the echo. High-tech or low-tech, the figure comes out about the same.

Skipping past the Dark Ages, dominated by Ptolemy's brilliantly wrong cosmology, with all the heavens looping around our planet like an amusement park ride, the next great leap didn't come until the sixteenth century. Copernicus restored the sun to the center of the solar system, then Kepler refined the model, flattening the planets' circular orbits into ellipses and showing how they must be arranged.

One of his laws is of particular interest to any would-be measurer of the universe: The farther a planet is from the sun, the longer it takes to complete its journey. From the length of the Martian year, one could deduce that the planet lies about one and a half times farther from the sun than Earth does. The same calculation could be done for Mercury, Venus, Jupiter, Saturn. . . . Once you had all the ratios, you could use parallax to determine just one of the distances. The others would automatically follow.

The task was easier said than done. Measuring the tiny displacement of the moon, viewed from two spots on the Earth, had been difficult enough. For even the closest planets, the parallax was so small that the slightest mistake in gauging an angle or the length of a baseline caused the calculation to fail.

That didn't keep astronomers from trying. Coordinating their efforts with pendulum clocks, observers stationed in Paris and on the island of Cayenne in South America showed in 1672 that the position of Mars shifted by a mere 25 seconds of arc. Adopting the useful fiction that the heavens consist of a hemispherical dome arcing overhead, astronomers divide the sweep from horizon to horizon into 180 degrees, half a circle. Each degree is divided into 60 units called minutes, which can each be further subdivided into 60 seconds. Twenty-five arc-seconds is $1/144$ of a single degree, an exceedingly small piece of sky.

The Martian distance computed from this delicate operation came close to the mark, but the accuracy was accidental. There was so much uncertainty in the measurements that errors canceled out errors, dumb luck stumbling onto a pretty good answer. Before more reliable methods became possible, another hundred years would go by.

2

Twice a century, just a few years apart, Venus, the closest planet, passes between the Earth and the sun. The result is like an eclipse, except that Venus is so distant that it appears only as a tiny dot. No one would notice the event who wasn't deliberately watching. If observers in separate locations time how long it takes the planet to cross the solar disk, they can compare their different readings and triangulate. The result is Venus's distance and, plugging the number into Kepler's equations, the distances of every planet from the sun.

Edmond Halley, the great British astronomer, had missed his chance to observe this rare phenomenon, called the transit of Venus. He lived inconveniently between the pair of events of 1631 and 1639 and the recurrence predicted for 1761 and 1769. He satisfied himself by issuing a challenge, calling on the next generation of astronomers to disperse themselves around the world and measure the Venusian parallax.

They took him up on the dare, setting out for Siberia, Hudson Bay, Baja California, India, the Cape of Good Hope, Tahiti. Some of the expeditions failed, and some of the data was suspect because of the difficulty of determining precisely when fuzzy-edged Venus, a planet swathed in chemical clouds, actually passed into the circle of the sun. But with eyes placed all over the planet (there were 150 observations), astronomers compiled enough good information to measure all the way to Venus. Then, carrying the calculation a step further, they concluded, via Kepler, that the distance to the sun was 91 million miles, just 2 million shy of the number schoolchildren learn today.

It was a hard act to follow. With one star down and a whole galaxy to go, the craft of triangulation was already becoming stretched to the limit. Even with an imaginary triangle whose base extended across the full width of the Earth, the parallax of the nearest planets was just barely measurable. How could anyone hope to gauge the distance of stars?

Astronomers had made some rough estimates. If you pretend that all stars produce the same amount of light—that they are equal in intrinsic brightness—you can judge how much dimmer Sirius appears from Earth compared with our home star, the sun. Then with the inverse square law (something twice as far away shines a fourth as bright) you can estimate the distance. These rough calculations served to show that even the brightest stars must be hundreds of thousands of times farther than the sun, way beyond the reach of earthly parallax. Travel from one antipode to the other and you would not detect the slightest shift.

It seemed that measuring stellar distances would require leaving the planet, observing the position of a star from Earth and then from a point millions of miles away. . . . Or you could stay on Earth and let it carry you between the far extremes of its orbit. This was the next advance. The radius of this great ellipse—the distance from Earth to sun—was now known with some precision. So just double the amount: every six months earthlings are looking at the sky from positions separated by

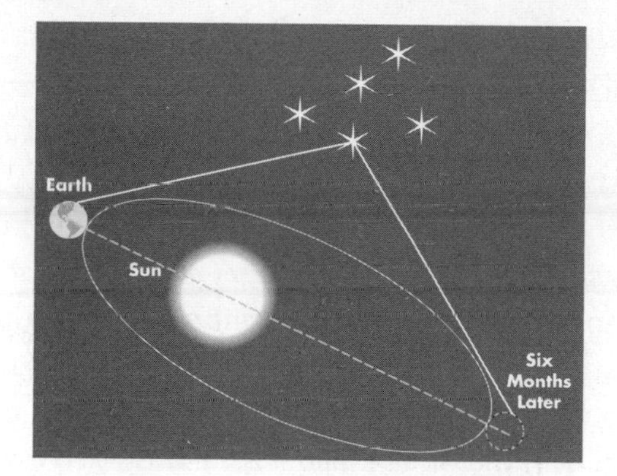

Parallax using the diameter of Earth's orbit as a baseline

186 million miles. Draw a triangle with this as the baseline and you can measure the parallax of nearby stars.

Galileo had suggested how such an experiment might be done. The sky is filled with double stars, some of which are presumably optical illusions: nowhere near each other in space, they just happen to line up because of the angle at which we see them. If one is actually much closer to us than the other, parallax will make them appear to come together and then move apart as we orbit the sun. (Think of two telephone poles, one behind the other, converging and then diverging as you drive by.)

It wasn't until the late 1700s that instruments were good enough to make the measurements. The astronomer William Herschel built a 20-foot-long telescope with a mirror 19 inches across. When that wasn't strong enough for him, he built a 40-foot scope with a 4-foot mirror—so large that it weighed a ton. Working with his sister, Caroline, he discovered Uranus and some 2,000 star clusters and nebulae (which, he ventured, were not small, nearby gas clouds but galaxies so distant that the individual stars blurred together). He also found hundreds of the double stars Galileo had suggested might be used for measuring distance.

In the end the project failed. A statistical survey showed that it was actually very rare for two stars to line up and mimic a double. Most doubles really were doubles, adjacent stars orbiting each other—too close together to exhibit any parallax.

It was with the next generation of astronomers that stellar triangulation came of age. Herschel's son John (the one we heard earlier rhapsodizing about the Magellanic Clouds) established an observatory at the Cape of Good Hope, near the southern tip of Africa. There astronomers determined that every six months the star Alpha Centauri changed in position by less than a second of arc—1⁄10,000 of a single degree, a frustratingly tiny amount but enough to do some trigonometry. Calculating the height of this extremely skinny triangle led to the conclusion that the star was 25 trillion miles away—so far that its light took more than four years to get here. And this was just the next sun over, the closest star.

Two other neighbors were also measured around this time, Vega and a star called 61 Cygni. A little later came Sirius and Procyon. All were a few light-years away in a universe now believed to be billions of light-years wide. By the early 1900s, when Henrietta Swan Leavitt arrived at Observatory Hill, nearly a hundred more stars had been triangulated. But most stars by far showed no parallax at all, even across so enormous a baseline. They were inconceivably and, it seemed, immeasurably far away.

3

Had just one of Miss Leavitt's blinking stars been within triangulating distance of home, astronomers could have leapt past the parallax barrier and begun measuring deep into space. Remember that two of her Cepheid variables pulsing at the same rate are, according to the relationship she discovered, of the same inherent brightness. If one appears to shine only a hundredth as brightly as the other, you know (because of the inverse square law) that it is ten times farther away. If you could use parallax to establish the distance of just one nearby Cepheid, you could infer the distance to the rest. By comparing Cepheid variables of various rhythms and intensities, you could leapfrog across the universe.

Nature, however, was not so obliging. The nearest known Cepheid, the North Star, was too far to show any shift in position, even when viewed from opposite ends of Earth's orbit. It is, by modern reckoning, some 400 light-years away. Parallax would get you only a fraction that far. The Cepheids used to devise Leavitt's law were many times farther still.

The best an astronomer could do was to say that a certain Cepheid was, as determined by its rhythm, one-tenth the distance of those in the Small Magellanic Cloud, while another was, perhaps, three times farther away. The universe could be portioned out in units called SMCs. But that begged the question: How far—in miles or light-years—was the Small Magellanic Cloud?

Before Miss Leavitt's stars could be turned into true yardsticks, some way had to be found to extend parallax far enough to reach a nearby Cepheid. That meant making observations across an even longer baseline than the full width of spaceship

Earth's orbit around the sun. What was needed was a larger, faster craft. Unlikely as it seemed, one was at hand: the starship we call the sun.

According to folklore, when Galileo was called before the Inquisition and forced to recant his Copernican views, he muttered under his breath, "And yet it moves." The reference, of course, is to our planet. He might have been as surprised as his tormentors to learn that the sun moves as well, on a slow drift through the Milky Way, carrying its planets alongside.

The movement is barely perceptible. In the late 1700s the elder Herschel, William, discovered that stars in the direction of the constellation Hercules move according to a peculiar pattern: over the years, they appear to be fanning out from a distant point, as snowflakes seem to do when viewed in the headlights of a car speeding through the night. In the opposite direction, back toward the constellation Columba, the stars converge, like snowflakes seen from the rear window. Our solar system, he concluded, was leaving Columba and heading toward Hercules. Astronomers have since clocked the speed of this journey through the galaxy at about 12 miles per second, or more than 30 million miles a year. The parallax from the voyage causes the constellations to become stretched and squeezed over time. The ancient Greeks were looking at a slightly different sky.

For a measurer of the universe, riding along with the sun, closer stars will move faster than farther ones, while the most remote stars will not seem to move at all. Carefully note the position of a Cepheid, then measure it again years later, when the sun has dragged Earth and its astronomers to a new location in space. Calculate the length of this enormous baseline, and then triangulate. With the distance of one Cepheid established, you can calibrate Leavitt's yardstick and then measure the rest.

The first to attempt this was a Danish astronomer named Ejnar Hertzsprung. He used the motion of the sun to triangulate the distance to some Cepheids in the Milky Way. Then, correlating pulse rate with inherent brightness, he extrapolated outward, reporting in a journal called *Astronomische Nachrichten* that the distance of the Small Magellanic Cloud was about 3,000 light-years.

This was enormous by the astronomical standards of the time. In fact it was a misprint. Maybe a journal copy editor had recoiled at the real number, inadvertently dropping a zero. According to Hertzsprung's calculations, the nebula was ten times farther, 30,000 light-years away.

Around the same time the American astronomer Henry Norris Russell used a different method to come up with an even more astonishing distance of 80,000 light-years. "I had not thought of making the very pretty use you make of Miss Leavitt's discovery about the relation between period and absolute brightness," he later wrote to Hertzsprung. "There is of course a certain element of uncertainty about this, but I think it is a legitimate hypothesis."

The Cepheid yardstick still needed refinement, but astronomers finally had hope of leaping past the nearest stars, roughing out the shape and size of the galaxy . . . and what, if anything, lay beyond.

4

Henrietta Leavitt didn't get to pursue the matter herself. Pickering kept her tied down with other projects. He wasn't one to encourage theorizing, believing, as his colleague Solon Bailey put it, "that the best service he could render to astronomy was the accumulation of facts." Beginning in August 1912, the year her discovery about the Cepheids was published, she documented her day-to-day routine, in language meaningful only to an astronomer, in a black-and-red leather-bound notebook:

> *October 8. Letter from Hertzsprung, dated Mount Wilson, Oct. 3, 1912. Subject, Method of transforming photographic to visual magnitudes by means of effective wavelengths. He finds that a change in the color index, using Dr. Mürch's photographic magnitudes and Harvard photometric magnitudes, of one magnitude, corresponds to a change in effective wavelength (λ_f) of 200 Angstrom units. . . .*
>
> *October 19. Tried superimposing Plates H 361, exp. 10m, limiting magn. 15.6 and H 385, isochromatic, limiting magnitude 14.9. The red stars appeared of nearly the same brightness on the two plates, white stars being brighter on H 361. The colors were assigned very easily.*
>
> *October 22. Finished revision of 32 sequences north of +75 degrees, and comparison with marked chart. Gave plates to Miss O'Reilly for identification.*

And so it went for the next four years, except for gaps, sometimes many months long, hints of recurring illness. In spring 1913 she was absent for three months recovering from stomach surgery. Only once in all those pages, on January 13, 1914, does she let some excitement show: "Completed discussion of Prov. Photovisual Magn. of N.P. Seq., this completing H.A. 71, 3. !!! after many years."

Translation: She had finished, or so she believed, the measurements for her North Polar Sequence, ninety-six stars whose magnitude she had determined with such authority and care that they could be used as a standard for the rest of the sky. There were still revisions to be done. The work was finally published, three years later, in the *Annals of the Astronomical Observatory of Harvard College*, volume 71, number 3—all 184 pages of it. Perhaps she and her mother indulged in a split of champagne.

Dry as dust to the uninitiated, her report was a work of magnificence, combining data from 299 photographic plates taken by thirteen different telescopes. Every magnitude had to be meticulously checked and cross-checked, with a constant awareness of the difference between mere data and true phenomena. No matter how carefully measured, each number represented not the brightness of a star but rather the intensity of its image on a photographic plate. In a perfect world these two values would be the same. In reality, every telescope and every type of photographic plate had its own peculiarities—responding more readily to some colors than to others. Images near the center of the plate were rendered more accurately than those to the side.

Page after page, she described how she corrected for the various biases and uncertainties. There was a chain of reasoning behind every number. Each star was a project in itself.

When she reached the end of the study, she knew it wasn't perfect. "It is desirable that the standard scale should be investigated by different observers, using independent methods," she allowed. "Discrepancies will inevitably appear in the results." But, as politely as possible, she warned future critics to proceed with caution.

"Too much weight may easily be assigned to results obtained from a single investigation, even if great precautions have been used." Which is why, she gently reminded, her measurements "depend on many different methods, instruments, and observers."

She concluded, "In view of these facts, it seems only reasonable that considerable time should be allowed to pass, and a large amount of varied material collected, before adopting definitive corrections to the scale here presented. For stars between the tenth and sixteenth magnitudes, such corrections are likely to be minute. For brighter and fainter stars, sensible changes may be made ultimately, but the scale is probably a close approximation to the true one."

It was work to take pride in. Ph.D.s have been awarded for less.

CHAPTER 5

Shapley's Ants

Her discovery of the relation of period to brightness is destined to be one of the most significant results of stellar astronomy, I believe. I am quite anxious to have her opinion as to the periods because of its bearing on some statistical work I am now bringing to a close.

—*Harlow Shapley, writing to Edward Pickering about Henrietta Leavitt*

Back in the eighteenth century, William Herschel had theorized that nebulae, like Andromeda, might be distant galaxies. The philosopher Immanuel Kant, called them "island universes," arguing, "It is much more natural and reasonable to assume that a nebula is not a unique and solitary sun, but a system of numerous suns." The universe, he wrote, may be filled with Milky Ways.

Other astronomers, however, were swayed by a different philosopher, Pierre-Simon Laplace, who proposed that the spiral-shaped clouds like Andromeda were no more than "proto solar systems"—a new sun and its orbiting planets in the process of congealing from a whirling cloud of gas. The theory seemed all the more plausible when in 1885 a new star, or "nova," flared within the center of the hazy disk of Andromeda, as though a solar system was being born.

Time exposures of the sky soon revealed some 100,000 of

the luminous swirls and more were being found all the time. With spiral nebulae appearing everywhere, it seemed absurd to suppose that each could be a galaxy filled with millions of stars. "No competent thinker, with the whole of the available evidence before him, can now, it is safe to say, maintain any single nebula to be a star system of coordinate rank with the Milky Way," Agnes Clerke, an astronomer and historian of science, wrote in 1890. "A practical certainty has been attained that the entire contents, stellar and nebular, of the sphere belong to one mighty aggregation." The universe was just another name for the Milky Way.

But the issue was far from settled. For one thing, the Milky Way itself seemed to be shaped like a spiral. Viewed from afar it might appear little different than Andromeda or any other nebula. An even stronger argument for island universes came from analyzing a nebula's light with a prism, breaking it into its component colors. These spectroscopic patterns were believed to reveal which chemicals a celestial object was made from. Andromeda's rainbow looked very much like the one cast by the sun. Both seemed to consist of star stuff. The tests were less than conclusive. Other nebulae produced dull, simple spectra—what one might expect from a homogeneous cloud of luminous gas. People tended to find in the data what they were already disposed to believe.

The impasse was broken in 1914 by Vesto Melvin Slipher at Lowell Observatory in northern Arizona, who had figured out how to estimate the speed at which a nebula was traveling through space. His technique was based on the Doppler effect. If a star is moving toward you, its light waves will be compressed. Thus the frequency—the number of waves that strike the eye every second—will increase. Brains interpret frequency as color, and so the starlight will shift toward the higher, bluer end of the spectrum. Conversely, if the star is moving away, its light will be stretched toward the lower-frequency reds.

Measuring the red and blue shifts of fifteen spiral nebulae, Slipher found them to be traveling at incredible velocities. Two of them appeared to be flying off at a dizzying 1,100 kilometers per second. That hardly seemed possible if they were simply small objects within the gravitational grip of the Milky Way. For many astronomers that settled the matter, in favor of island universes. (Slipher himself initially clung to the prevailing notion that a nebula was but a single star "enveloped and beclouded by fragmentary and disintegrated matter.")

More evidence arrived three years later when another nova suddenly appeared inside a spiral nebula called NGC 6946 (after its designation in the New General Catalog of Nebulae and Star Clusters). Heber Curtis of Lick Observatory, a California stronghold of the island universe theory, found novae inside other nebulae, and when astronomers reexamined old photographic plates they found still more.

Curtis believed the novae might serve as standard candles. Astronomers had estimated that those in the Milky Way surged to an intrinsic magnitude of about –8. (Remember that the lower the magnitude, the brighter the star, making one with a negative value very bright indeed.) Curtis assumed that the novae in the distant nebulae were probably peaking at about the same intensity. Compare that number with how bright a nova appeared from Earth and you could use the inverse square law to get its distance. Measured this way, the nebulae appeared to be huge spinning galaxies many millions of light-years away.

By 1917 the consensus had shifted toward island universes. In addition to Curtis, and by now Slipher, supporters included such prominent astronomers as Arthur Eddington, James Jeans, Ejnar Hertzsprung, and a young researcher named Harlow Shapley, who had recently moved to Mount Wilson Observatory, the astronomical powerhouse perched high above Pasadena, California. Shapley however was about to change his mind. Using Leavitt's variable stars, he would spend the next few years calibrating the Cepheid yardstick and measuring the size and shape of the Milky Way. He was ultimately forced to conclude that it was far larger than anyone had dared imagine—so large, he believed, that it must constitute the entire universe, nebulae and all.

2

Shapley had a peculiar fixation with ants. When he wasn't looking upward at the stars, he liked to watch a colony of medium-size brownish black ants—*Liometopum apiculatum*—as they streamed along a concrete wall by Mount Wilson's maintenance shop. Shapley noticed that the ants slowed down when they reached the shade of some manzanita bushes and sped up again in the sun. Armed with various instruments, he studied the ants under different atmospheric conditions, even watching them with a flashlight at night. The correlation between running speed and temperature was so tight that he could use the ants as a thermometer, reading off the temperature within one degree. Returning to Mount Wilson thirty years later, he was annoyed to find that an assistant engineer was making a practice of burning off the ant trail with a blow torch—"genocide," Shapley called it. ("Formicide" would have been a better word.) But the resilient ants always came back.

He drew lessons from these tiny creatures. Asked to deliver

a commencement address at the University of Pennsylvania, he chose the topic "On Running in Trails," warning the students against the comfortable allure of conformity, of following the same narrow paths as their ancestors, afraid to break from the pack.

When Shapley arrived in Pasadena in 1914, the common wisdom held that the Milky Way was a lens-shaped disk some 25,000 light-years long and about a fourth that wide, with the sun at almost dead center. This picture of the heavens was sometimes called the Kapteyn universe, after the Dutch astronomer Jacobus Kapteyn, who had estimated its size. The methods he had used were far from exact. With the Cepheid variables and Mount Wilson's powerful 60-inch telescope at his command, Shapley decided to measure the galaxy for himself.

Spread throughout the Milky Way were a hundred or so "globular clusters," each consisting of hundreds of thousands, even millions, of stars. Shapley suspected that these huge concentrations formed a kind of framework or "skeleton" mark-

The Milky Way

ing the extent and shape of the galaxy. By using Cepheids to determine the distances to these mileposts, he could map out the whole thing.

Shapley figured he knew something about variable stars. His Ph.D. dissertation at Princeton, under Henry Norris Russell, had focused on a type of variable called eclipsing binaries—two stars orbiting around a common point and periodically blocking each other's light. One of Shapley's first papers at Mount Wilson showed that the Cepheids did not belong to this class. Rather, they were single stars that expanded and contracted with a regular beat. For now, however, these details were unimportant. He knew from Henrietta Leavitt's research that Cepheids would serve as standard candles.

He also knew that most of the variables in the globular clusters were a little different from the ones she had discovered in the Magellanic Clouds. Shapley's stars—called cluster variables—blinked much faster, with cycles measured in hours, not days. Hertzsprung, in fact, thought his eager colleague was mixing apples and oranges. How could he be so sure that both kinds of stars showed the same connection between period and brightness?

Shapley was insistent. "[T]his proposition scarcely needs proof," he wrote in a paper in the *Astrophysical Journal.* "Practically all writers on the subject are more or less inclined to accept this view."

Undeterred, he proceeded with his plan. For his early measurements, he relied on the fact that the farther something is from an observer, the more slowly it appears to move—think of a tiny jet plane inching across a windowpane. The speed at which a star is heading directly toward or away from Earth, its "radial velocity," can be clocked using redshift and blueshift.

But it is the "transverse velocity"—how fast the star is moving across the sky—that hints at how far away it is. It seemed sensible that, on average, stars in a cluster would move at the same velocity, whatever the direction. Drawing on a method called statistical parallax, Shapley used Doppler shifts to estimate the average velocity of a sampling of stars and compared that figure with how fast the stars *appeared* to be moving. That revealed their distance. Eleven of these were Cepheids, forming the anchor of what came to be called "Shapley's curve."

Taking a second leap, he extended the curve to include the far more common shorter-period variables. First he would find a cluster that had both types. The slow-paced Cepheids gave him an estimate of the cluster's distance, which he could then correlate with the period of the faster variables. Now the distance of clusters with only fast variables could be measured—provided that both kinds of stars really obeyed the same law.

New yardstick in hand, he gauged the distances to several of the nearest globular clusters. Then he ran into a wall. In most of the clusters, not a single blinking star could be found. He would have to extrapolate further, and that meant finding another kind of standard candle. It seemed sensible, he reasoned, that each cluster would consist of stars spanning the same range of magnitude. The brightest stars in cluster A, whose distance had been measured with his yardstick, would be about as intense as the brightest stars in cluster B, whose distance was unknown. If they appeared dimmer, it would be because they were farther. The inverse square law would reveal by just how much, extending the map a little more.

Many clusters, however, were so distant and so blurry that not even Mount Wilson's telescope could pick out a single star. And so came the final leap: one could take a cluster whose distance had been established by these other indirect methods and use the whole thing as a standard candle. The farthest clusters, Shapley reasoned, were probably as intrinsically big and brilliant as the nearest ones. By measuring how much smaller and fainter they appeared, he could judge their distances, and reach to the farthest edges of the galaxy.

As he followed this artful chain of assumptions, Henrietta Leavitt was living alone in a Cambridge rooming house, where she had moved after the death of her uncle Erasmus in 1916. She was still working for the observatory and occasionally Shapley wrote to Edward Pickering inquiring about the latest developments with her variable stars.

Shapley had noticed some very faint variables in the Magellanic Clouds and wondered if these might be similar to the ones he was using to plot out the Milky Way. "Does Miss Leavitt know if they have shorter periods, that is, are their periods shorter than one day, similar to cluster variables?" he wrote on August 27, 1917. "It may be her work has not progressed far enough to give a definite answer." He was hoping for some ammunition against those, like Hertzsprung, who continued to argue that the faster variables did not necessarily obey Leavitt's rule. He considered the matter "of much importance. . . . In fact, the Magellanic clouds and their variables seem to me one of the most important outstanding problems of stellar photometry."

Pickering replied about three weeks later: "Miss Leavitt is now absent on her vacation." (She was on Nantucket, visiting with Margaret Harwood, a fellow Harvard computer and astronomical assistant who had become director of the Maria Mitchell Observatory.) "When she returns, she will investigate the matter of the Magellanic Clouds."

Of the two men, Shapley was the quicker and more loquacious correspondent. Within a week he had fired off another letter, praising Leavitt's work and emphasizing how much he needed the information. "Her discovery of the relation of period to brightness is destined to be one of the most significant results of stellar astronomy, I believe. I am quite anxious to have her opinion as to the periods because of its bearing on some statistical work I am now bringing to a close."

Nine months later Shapley was still waiting. On July 20, 1918, he checked in again, still heaping on the praise:

> I believe the most important photometric work that can be done on Cepheid variables at the present time is a study of the Harvard plates of the Magellanic clouds. Probably Miss Leavitt's many other problems have interrupted and delayed her work on the variables of the clouds for the interval of six or seven years since her preliminary work was published. . . . The theory of stellar variation, the laws of stellar luminosities, the arrangement of objects throughout the whole galactic system, the structure of the clouds—all these problems will benefit directly or indirectly from a further knowledge of the Cepheid variables.

It took almost three weeks for Pickering to reply: "A few days ago I talked with Miss Leavitt. . . . She has the material for about a third of the brighter variables, and photographs are now being taken with the Bruce 24-inch, which I hope will provide the remainder."

That is the last letter from Pickering in Shapley's files. Less than five months later, he died of pneumonia at age seventy-two.

3

Shapley ultimately decided to run with his theory, using the Cepheids as the first step in his hopscotch across the galaxy. The results were astonishing. First of all the Milky Way, by Shapley's measure, was gargantuan in size—300,000 light-years across. That was some ten times greater than Kapteyn's estimate—so much larger that he felt he must abandon the notion of island universes. If one insisted on maintaining that the thousands upon thousands of spiral nebulae were galaxies each the size of the Milky Way, then Andromeda, judging from its apparent size, would have to lie at an enormous distance. That, in turn, would mean that its novae were absurdly bright. It must be a small gas cloud after all.

To Shapley there was now an even more damning argument against island universes. One of his colleagues, a Mount Wilson astronomer named Adriaan van Maanen, had recently announced that several of the great spirals, including the aptly named Whirlpool and Pinwheel nebulae, were gradually turning. Van Maanen made the measurements with an arrangement of lenses and mirrors called a blink comparator. With the device, an astronomer could mount two photographic plates taken months or years apart and, gazing through a binocular eyepiece, switch back and forth between them. Anything that had changed would appear to move or vary in size. Comparing plates of nebulae taken five years apart, van Maanen thought he saw a slight rotation.

Viewed from Earth the spin was minuscule—2/100 of a single second of arc each year. A complete cycle would take about 100,000 years. The surprise was that the movement was visible at all. Few could doubt that spirals spun. Why else would they have their pinwheel shapes? But for the motion to be perceptible at all, these nebulae would have to be small and nearby. If the spirals were truly distant galaxies, van Maanen's data would mean they were spinning at impossibly high velocities—faster than the speed of light.

Just as unsettling as the Milky Way's enormous size was Shapley's conclusion about where we lived inside the galactic disk. Astronomers had noticed that the globular clusters are not distributed evenly through the sky but congregate in the direction of the constellation Sagittarius. There, according to Shapley's measurements, they formed a roughly spherical shape, a cluster of clusters. That, he proposed, must be the center of the galaxy, the central bulge of the Milky Way. If we lived within this region, we would see the clusters spaced uniformly around us. The fact that we do not is because we lie in the galaxy's outskirts, tens of thousands of light-years from the core. We were not New York City but Pensacola or North Platte.

"So the center has shifted: egocentric, lococentric, geocentric, heliocentric," Shapley wrote to George Ellery Hale, Mount Wilson's director. Or, as he later put it: "Man is not such a big chicken. If man had been found in the center, it would look sort of natural. We could say, 'Naturally we are in the center because we are God's children.' But here was an indication that we were perhaps incidental. We did not amount to so much." People were no more important than ants.

The center of the galaxy was in Sagittarius. And so the center of the universe must be there as well.

CHAPTER 6

The Late, Great Milky Way

> The spectrum of the average spiral nebula is indistinguishable from that given by a star cluster. It is such a spectrum as would be expected from a vast congeries of stars.
>
> —*Heber Curtis*

On a spring day in 1920, Shapley found himself strolling along the train tracks somewhere in Alabama talking about flowers and classics and probably looking for ants. His companion was Heber Curtis, and they had agreed for now to avoid the touchy subject of astronomy. Each, unbeknownst to the other, had booked passage on the same train from California to Washington, D.C., where, in the halls of the National Academy of Sciences, they were scheduled to debate whether there was anything in the universe beyond the Milky Way.

The event had been arranged at the prompting of Shapley's boss, George Ellery Hale, one of the most revered astronomers of the time. Hale's father had made a fortune selling hydraulic elevators to the builders of Chicago's soaring skyscrapers. The son set his sights still higher, studying astronomy and exploiting his family's financial connections to fund the development of some of the best telescopes in the world. When the old man died, a series of lectures was endowed in his name. The

Harlow Shapley
(Harvard University Archives)

younger Hale thought that the one in 1920, at the National Academy's annual meeting, should be devoted to a currently hot cosmological issue—either relativity or island universes.

The first topic struck the academy's secretary as too esoteric. (He personally thought that Einstein's theory should be banished "to some region of space beyond the fourth dimension, from whence it may never return to plague us.") Nor was he keen on island universes, fearing that "unless the speakers took pains to make the subject very engaging the thing would fall flat." He proposed instead a presentation on glaciers or "some zoological or biological subject." In the end, Hale had the final word, and Shapley and Curtis were picked to present their opposing views on "The Scale of the Universe," and, more pointedly, on whether it consisted of more than a single galaxy.

Were such an event to take place today, it would be videotaped and perhaps transcribed. You could probably download it from the Web. The encounter between Shapley and Curtis can be pieced together only from scraps of evidence—a typescript of Shapley's talk, annotated with his scribbles, some of Curtis's slides (he misplaced his script soon after the event), and letters the two exchanged before and after what came to be called the Great Debate, mostly by people who had not been there.

Curtis himself was eager for a scrap. He imagined the two astronomers going after each other with "hammer and tongs," then shaking hands like gentlemen. Shapley, however, was worried that he might lose. Not that he thought his theory was wrong. But persuading an audience of geologists, biologists, and scientists of other nonastronomical persuasions required the skills of an orator. Right or wrong, Curtis, thirteen years older and the more polished debater, might best Shapley at the podium.

He expected for example that Curtis would pick on the tiny handful of stars ("my eleven miserable Cepheids," Shapley called them in a letter) from which he had extrapolated such enormous distances. What seemed to some like a brilliant analysis might strike others as a house of cards. Many astronomers were much less confident than Shapley about the usefulness of Henrietta Leavitt's yardstick. Curtis had made it clear that he thought Shapley's Milky Way was ten times too large. If he could shrink it back by an order of magnitude, the island universe theory might be easier to uphold.

Shapley also had an ulterior motive. He was certain he was being considered for the directorship of Harvard Observatory—Edward Pickering had just died—and he expected an emissary from Observatory Hill to be in the audience. He dreaded making a bad impression.

For weeks Shapley worked to bolster his evidence while maneuvering for a more advantageous position. Through sheer obstinacy, he managed to get the debate downgraded to a discussion—"two talks on the same subject from our different standpoints"—and the talks themselves reduced in length. While Curtis wanted forty-five minutes for each presentation, Shapley wanted thirty-five. They split the difference, settling on forty. To blunt the impact further, no time would be allowed for rebuttals, just a general discussion at the end. Finally Shapley made certain that Henry Norris Russell, his former teacher and ally, would be in the audience to support his position. He wasn't taking any chances.

2

The evening began at an excruciating pace with awards followed by long testimonials. There was a tribute to the Prince of Monaco for oceanography, Shapley later remembered, and another to some "noble human antique," honored for combating hookworm. Many years later in a memoir, Shapley recalled a bored Albert Einstein sitting in the audience, whispering to his companion that he now had a new theory of eternity. It made a good story, but actually Einstein was in Germany then, fending off the first stirrings of Nazi denunciations of his "Jewish physics." He made his first visit to the United States the following year.

Once the main event finally began, Shapley went first, easing in slowly with a long introductory tutorial on astronomy.

A third of the way through his allotted time, he had gotten only as far as the definition of a light-year. So far the presentation was pure popular science. One can imagine Curtis glancing at his watch, wondering when Shapley would say something he could dispute.

Then he surprised Curtis again with a promise to spare the audience "the dreary technicalities of the methods of determining the distance of globular clusters," the way stations he had used to map the galaxy. Maybe that was a sensible approach for an audience of nonspecialists. But Curtis, who was still itching for a fight, had prepared a meticulous point-by-point deconstruction of Shapley's every assumption and logical inference. There was still nothing for him to rebut.

The biggest surprise was still to come. Skipping over Cepheids entirely, Shapley described an entirely independent method of establishing the enormity of the galaxy (and by implication, undermining the case for island universes).

Astronomers had uncovered what appeared to be a relationship between a star's temperature and its inherent brightness. (These had been plotted on a chart famous to astronomers as the Hertzsprung-Russell diagram after Henry Norris Russell and Ejnar Hertzsprung.) The result was another kind of measuring stick. A survey of nearby "B-type stars," identified by their bluish sheen, had found them to be on average 200 times brighter than the sun. This suggested, to Shapley anyway, that these blue giants could be used as standard candles. No matter how much a B star's light is dimmed on the journey to Earth, it could be assumed, with a small leap of faith, to have the same intrinsic brightness as its cousins closer to home. And so the giant's distance could be calculated from the inverse square law. If it was nine times dimmer than a nearby B star, it must be three times farther away.

Shapley found what he took to be blue giants in the Milky Way cluster called Hercules. Exceedingly dim—of the fifteenth magnitude—the stars, he reckoned, would have to be 35,000 light-years away. Then he extrapolated further. Assuming that the smaller, dimmer clusters had the same overall brightness as Hercules, he circled around to his original conclusion: that they lie at the fringes of a galaxy 300,000 light-years in diameter, with the sun shoved off to one side.

He briefly mentioned how another kind of yardstick, stars called red giants, also supported his measurements. Then he tried to fend off any criticism of his "miserable" Cepheids by removing them from the debate: Professor Curtis, he said, "may question the sufficiency of the data or the accuracy of the methods. . . . But this fact remains: we could discard the Cepheids altogether, use instead the thousands of B-type stars upon which the most capable stellar astronomers have worked for years, and derive just the same distance for the Hercules cluster, and for the other clusters, and obtain consequently the same dimensions for the galactic system."

Before the debate the red and blue stars had been no more than footnotes to Shapley's argument—secondary checks on his primary measuring tool, Henrietta Leavitt's Cepheid variables. Now figure and ground had been reversed, leaving Curtis to aim at a moving target and leading one to wonder who really was the wilier debater.

As for the nature of the spiral nebulae, the original focus of the discussion, Shapley dismissed them in a few sentences:

> I shall leave the description and discussion of this debatable question to Professor Curtis. We agree, I believe, that if the galactic system is as large as I maintain, the spiral nebulae can hardly be comparable galactic systems; if it is but one-

> tenth as large, there might be a good opportunity for the hypothesis that our galactic system is a spiral nebula, comparable in size with the other spiral nebulae, all of which would then be "island" universes of stars. On one other point I think we also agree, or at least we *should* agree, and that is that we know relatively so little concerning the spiral nebulae . . . that it is professionally and scientifically unwise to take any very positive view in the matter just now.

Even if the spiral nebulae were not firmly inside the boundaries of the Milky Way, he believed, they probably lay on the outskirts, small gas clouds encountered during the galaxy's drift through endless nothingness.

3

There is no way to re-create the tone of Curtis's presentation from the skeletal talking points left behind on his typewritten slides. It is clear that, undeterred by Shapley, he marshaled a strong defense of island universes. As expected, he challenged the reliability of Shapley's calibration of the Cepheid yardstick, joining those who suspected that the more rapidly oscillating cluster variables used to measure out the Milky Way were different in nature from the slower ones Leavitt had found in the Magellanic Clouds. Why assume they had precisely the same relationship between period and inherent brightness, if there was indeed any such relationship at all? If Shapley's variables were actually much dimmer to begin with, then all the clusters would lie closer in. The perimeter of the galaxy would contract and the Milky Way would shrink from continent to island, one member of a great archipelago.

Heber Curtis
(Lick Observatory)

Curtis was equally unimpressed with the giant blue stars, arguing that far too little was known to trust them as standard candles. He proposed what he considered a more reliable measuring device—the yellow-white stars like the sun that seemed to make up most of the galaxy. It was reasonable, Curtis proposed, that the sunlike stars in the far reaches of the Milky Way shine, on average, with the same brightness as those nearby. Like Shapley, he was assuming the uniformity of nature, and his conclusion was that the galaxy can be only about 30,000 light-years across. For the clusters to be as distant as Shapley believed, these stars would have to be far brighter than those in our own neighborhood. A different physics would prevail. "While it is not impossible that the clusters are exceptional regions of space [with] a unique concentration of giant stars, the hypothesis that cluster stars are, on the whole, like those of known distance seems inherently the more probable."

With the Milky Way knocked down in size, evidence for island universes seemed compelling. Curtis recalled the familiar argument that the color patterns produced by the spirals were indistinguishable from that of starlight. "It is such a spectrum as would be expected from a vast congeries of stars." And the novae that appeared within them "seem a natural consequence of their nature as galaxies, incubators of new stars." Used as standard candles the novae put Andromeda half a million light-years from Earth and other spirals 10 million or more light-years away. "At such distances, these island universes would be of the order of size of our own Galaxy of stars."

Finally he noted a curious phenomenon that had been puzzling astronomers for years: the spiral nebulae appeared to be concentrated at the two "poles" of the Milky Way—the regions directly above and below the galaxy's central bulge. None was found in the galactic plane where most of the stars reside. If the spirals were small clouds within or near our galaxy, then why were they not evenly distributed? It was as though they were being repelled by some mysterious force.

It was far more plausible, Curtis argued, that this "zone of avoidance" was an illusion—that the spirals lay far beyond the Milky Way, in every direction, with those along the galactic plane hidden from our view. Many spirals appeared to be surrounded by a thick ring of "occulting" matter, a halo of interstellar dust. The same might be true of the Milky Way. When astronomers aimed their telescopes directly into this dust storm, spirals in that direction were blocked from view. Of the millions of spirals, the only ones we could see were those that happened to lie above and below. Each, he proposed, was a world as vast and shining as our own.

4

Each man left the lecture hall certain that he had won. "Debate went off fine in Washington," Curtis wrote to his family, "and I have been assured that I came out considerably in front." Shapley, for his part, attributed any points Curtis may have scored to his rhetorical skills. "Now I would know how to dodge things a little better," he said years later, a comment that seems strange since Shapley spoke first. Maybe he was referring to the discussion that followed the talks, during which his mentor Russell had, as planned, come forth with a strong endorsement of Shapley's Big Galaxy theory. The champions of island universes surely responded just as vigorously. "Curtis did a moderately good job," Shapley recalled. "Some of his science was wrong, but his delivery was all right."

Both men fleshed out their arguments in papers published the following year in the *Bulletin of the National Research Council.* (In some early historical accounts these published papers are treated as the actual substance of the debate.) The texts contain no fundamentally new arguments. Shapley bolstered his case with more data, whose certainty Curtis continued to question. What is most striking is how two of the world's smartest astronomers could take the same trove of astronomical observations and come up with two such very different pictures of the universe, a reminder that science lies not in the facts themselves but in their arrangement.

For Curtis, the zone of avoidance (it sounds like something

from a Superman comic) was strong evidence for seeing the spirals as island galaxies. In Shapley's hands, the phenomenon seemed to support the argument that spirals are little wisps of stellar gas: they would have to be small and light to be repulsed somehow by the Milky Way. Also open to conflicting interpretations were the novae that appeared now and then inside the spirals. For Curtis their existence showed that spirals were indeed galaxies. Where else would one expect to observe stars being born? For Shapley each nova represented "the engulfing of a star by [a] rapidly moving nebulosity."

The most memorable passage in either paper is a paragraph by Shapley on the perils of extrapolating too boldly from a limited set of data. He meant the statement as a criticism of Curtis's measurements involving the average magnitude of sunlike stars. But it could be taken just as easily as a humbling reminder about how much the entire enterprise of astronomical measurement rests on a few vulnerable assumptions. (It is also the inspiration for the story about the villagers in the canyon, which appears in the prologue of this book.)

"Suppose," Shapley began, "that an observer, confined to a small area in a valley, attempts to measure the distances of surrounding mountain peaks." He can use parallax for the nearby hills, but since he cannot leave his narrow valley, his baseline is too small to triangulate any farther. He needs another kind of measuring stick. Seeing through his telescope that there is plant life on the mountaintop, he makes the simplifying assumption that it is approximately the same as the plant life on the valley floor—averaging about a foot in height. Thus from the apparent size of the foliage, he can judge how far the mountain is.

His calculation would be wrong. "If, however," Shapley noted, "he had compared the foliage on the nearby, trigonometrically-measured hills with that on the remote peaks, or had used some method of distinguishing various floral types, he would not have mistaken pines for asters and obtained erroneous results for the distances of the surrounding mountains. All the principles involved in the botanical parallax of a mountain peak have their analogues in the photometric parallax of a globular cluster."

Mistaking pines for asters, and asters for pines. It was an occupational hazard that would plague Shapley as much as anyone.

CHAPTER 7
In the Realm of the Nebulae

One of the few decent things I have done was to call on her on her death bed. It made life so much different, friends said, that the director came to see her.

—*Harlow Shapley, writing about Henrietta Swan Leavitt*

For two boys hailing from rural Missouri, Harlow Shapley and Edwin Hubble didn't have much in common, except perhaps the size of their egos. Shapley had been born in 1885 on a hay farm near the Ozarks and dropped out of school to become a police beat reporter for a small-town newspaper. He had completed the equivalent of the fifth grade. Only a few years later did he earn a high school diploma and enroll at the University of Missouri. There he was diverted from journalism into astronomy, finally moving on to Princeton to study under Russell, who once told Edward Pickering, "He is the best student I ever had." For all Shapley's self-assured swagger—he was certain that he had mapped the length and breadth of the universe—he never quite lost his rough country edge.

Four years later, and about seventy miles east of the Shapley farm, Hubble was born. The family was more prosperous than the Shapleys (Mr. Hubble was an attorney turned insurance executive). A straight-A student and an athlete, Edwin won a

Edwin Hubble (Courtesy of the Archives, California Institute of Technology)

scholarship to the University of Chicago and became a Rhodes Scholar at Oxford, where he studied law and picked up the fake British accent that would grate on some of his colleagues like a kid squeaking a balloon.

Hubble, like his father, didn't take to practicing law. After briefly working as a high school teacher in Indiana (appearing before his classes in knickers and a cape), he returned to Chicago to pursue a doctorate in astronomy. Then, after serving as an officer in World War I, he came to Mount Wilson, where he and Shapley found themselves working uncomfortably under the same dome-shaped roof. What, Shapley wondered, was a Missouri boy doing exclaiming, "Bah Jove," or remarking that a plan had "come a cropper"? Taking note of Hubble's aristocratic and somewhat military bearing, some astronomers began referring to him as the Major.

Hubble, a rather reserved sort, found Shapley overbearing and erratic—shooting off one wild idea after the other—and he was particularly put off by Shapley's friend, the garrulous Dutch astronomer Adriaan van Maanen, whose love for dinner parties and socializing made him something of a standout in stodgy Pasadena. Van Maanen was also known as a meticulous astronomer—his measurements of the rotational velocity of the spiral nebulae provided the strongest argument against island universes. Having rechecked and refined his data, he remained adamant: either the spirals were small and nearby or they were spinning at insanely high speeds. He believed, as his teacher Kapteyn had taught him, that there can be only one galaxy, the Milky Way.

Hubble himself leaned toward the island universe theory, but for now he wasn't involving himself in the controversy. He hadn't come to Mount Wilson to confer with mortals about their astronomical opinions. The answers would be found in the stars. Within weeks he was sitting through long nights beneath the observatory dome, consorting with the sky. He spent Christmas Eve 1919 with his eye to Mount Wilson's new 100-inch telescope, 40 inches greater in diameter than the one Shapley had used to map out his Big Galaxy. For the next three decades it would be the largest in the world. Hubble was looking particularly hard at nebulae and wondering, perhaps, when Harlow Shapley was going to get off his mountain.

2

By now, Henrietta Leavitt and her widowed mother had taken up housekeeping in a recently built brick apartment building on Linnean Street and Massachusetts Avenue, several blocks from Harvard Observatory. Although she was largely occupied with more routine tasks, variable stars were still very much on her mind. In 1920 she wrote to Shapley seeking his advice. Where would he suggest she focus her research next? He replied, still hammering on an old theme, that it would be of "enormous importance in the present discussion of the distances of globular clusters and the size of the galactic system" if she would plot the periods of some of the dimmer variables in the Small Magellanic Cloud, those "just fainter than the faintest already studied." This is what he had been pestering Pickering about until several months before his death.

And perhaps she would see if her discovery about the Cepheids also held for those in the Large Magellanic Cloud. "Does the period-luminosity law apply there?" He was treating her almost like a colleague. Soon he would be her boss.

Shapley had been overestimating how badly Harvard wanted him. At first the university's president, Abbott Lawrence Lowell, had considered him the obvious choice to succeed Edward Pickering as director. But after consulting with several astronomers, Lowell found himself leaning toward offering Shapley the number-two spot, with an older, more experienced astronomer like Henry Norris Russell running the show. Shapley struck some of his colleagues as young and immature, and perhaps too brash for Harvard. "He is much more venturesome than other members of our staff," Shapley's boss, George Ellery Hale, confided to Lowell, "and more willing to base far-reaching conclusions on rather slender data."

And, as Shapley had feared, his lackluster performance at the Great Debate hadn't helped matters. Even Russell came away persuaded that his protégé was not ready to run Harvard Observatory. "Shapley couldn't swing the thing alone," Russell told Hale. "I am convinced of that after trying to measure myself with the job, and observing Shapley in Washington. But he would make a bully second."

In the end Shapley, willful as ever, got what he wanted. Ultimately Russell turned down Harvard's offer and Shapley made it clear that he wouldn't settle for anything less than the directorship. Hale interceded on his behalf and Harvard agreed to try out the young astronomer on a one-year probation. In the spring of 1921 he moved to Cambridge to take over where Pickering had left off.

"MARCH 28, 1921: Dr. Shapley arrived!" wrote Annie Cannon, who had become one of the most accomplished of the observatory's women assistants, in her diary. "I like him. So young, so clean, so brilliant." Like Henrietta Leavitt, Cannon was deaf, though only partially so. The following week she and a friend invited Shapley over for dinner and they all went to the symphony.

By now Leavitt was the head of stellar photometry, and the ebullient Cannon was curator of the photographic plate collection and chief compiler and overseer of the Henry Draper Catalogue of stellar spectra. Ultimately it filled nine volumes with more than 225,000 stars classified according to their spectral type, from the hottest white-blue stars to the cooler yellow, orange, and red. (Cannon's categories were called, cryptically, O, B, A, F, G, K, and M, which astronomers, some of the men anyway, remember with this mnemonic: "O Be A Fine Girl, Kiss Me").

Under Pickering, the status of the women computers had

continued its slow climb. He even tried, with no success whatsoever, to persuade the president of Harvard to grant Cannon the prestige of an academic appointment, or at least to include her name in the school's catalog. Women were praised, a little condescendingly, for being good at detail work, the numerical needlepoint of analyzing astronomical imagery, but deeper matters were still reserved for the men. One of the more overqualified assistants, Antonia Maury, a Vassar graduate, chafed under the tedium. "I always wanted to learn the calculus," she later said, "but Professor Pickering did not wish it."

Maury, truth be told, could be a pain to work with. She had been hired because she was Henry Draper's niece. Her work was slow and erratic, prompting her aunt to apologize for her behavior. "I shall be happy," she had written to Pickering, "when you are rid of the annoyance." But Maury's bad attitude was inflamed by a feeling that she was being discouraged from making original contributions.

In her own diary, Williamina Fleming, the housekeeper turned astronomical assistant, expressed the sense of frustration and stoicism some of the computers felt: "If one could only go on and on with original work, looking to new stars, variables, classifying spectra and studying their peculiarities and changes, life would be a most beautiful dream; but you come down to its realities when you have to put all that is most interesting to you aside, in order to use most of your available time preparing the work of others for publication. However, whatsoever thou puttest thy hand to, do it well."

Cannon was happier with her lot. When a young British astronomer named Cecilia Payne arrived to study at the observatory in 1923, she wondered how Cannon could have spent all those years under Pickering meticulously classifying stars without speculating on what the new taxonomy might mean.

In Cannon's case, Payne came to conclude, theorizing was against her nature: "She was a pure observer, she did not interpret." And she seemed to rely less on reason than on instinct. "She was like a person with a phenomenal memory for faces," Payne observed. "She did not think about the spectra as she classified them—she simply recognized them." When she needed to concentrate, she would disable her hearing aid.

Leavitt's feelings about her own work have gone unrecorded. No revealing confessions or letters have been found, just the occasional anecdote. One day, confronted with a particularly mysterious variable called Beta Lyrae, she exclaimed to a colleague, "We shall never understand it until we find a way to send up a net and *fetch the thing down*!" She yearned, perhaps, to rise above the columns of numbers and really know the stars. Yet even after her discovery of the Cepheid law, she remained assigned to routine photometry, more astronomical needlework. As Cecilia Payne later put it, "Pickering chose his staff to work, not to think."

Perhaps this would have changed with Shapley. More than anyone, he had seized on Henrietta Leavitt's stars to plumb the depths of space. He later called her "one of the most important women ever to touch astronomy." Considering how very few women there were in the field, it is hard to gauge how this weighed on his scales of praise. This was a man who measured the computational difficulty of astronomical jobs in "girl-hours" and the really difficult ones in "kilo-girl-hours."

Any chance the two might have had to collaborate was short-lived. Leavitt, still living with her mother on Linnean Street, was sick again, this time with cancer.

"Took flowers to Miss Leavitt who is very ill," Cannon wrote in her diary for November 6, 1921. It was a dreary month. By

Thanksgiving, Cambridge was besieged by the worst ice storm in memory. Trees and electric poles were breaking under the clinging sleet. The observatory lights went out.

Cannon's diary describes what came next:

December 6. Went to see poor Henrietta Leavitt, dying with a malignant stomach trouble. So thin & changed. Very, very, sad.

December 8. Clear and cold.

Shapley dropped by to pay his respects. "One of the few decent things I have done was to call on her on her death bed," he later said. "It made life so much different, friends said, that the director came to see her." Maybe so.

December 12. Rainy day pouring at night. Henrietta passed away at 10.30 P.M.

December 13. Mr. Leavitt, Henrietta's brother, called early in morning. Snowy, sloppy, dark day.

December 14. Wednesday. Henrietta's funeral at Chapel of 1st Cong. Church 2 P.M. Coffin covered with flowers.

She was buried at Cambridge Cemetery (across from the better known Mount Auburn), in the Leavitt family plot. Sitting at the top of a gentle hill, the spot is marked by a tall hexagonal monument, on top of which (cradled on a draped marble pedestal) sits a globe. Her uncle Erasmus and his family are also buried there, along with other Leavitts. A plaque memorializing Henrietta and her two siblings who died so young, Mira and Roswell, is mounted directly below the continent of Australia. Off to one side, and more often visited, are the graves of Henry and William James.

A few days before her death Leavitt had written out her will in longhand, leaving her mother an estate of odds and ends:

Bookcase and books $5
Folding screen $1
Rug $40
Table $5
Chair $2
Desk $5
Table $5
Rug $20
Bureau $10
Bed-stead $15
Mattresses (two) $10
Chairs (two) $2
One @ $100 face value First convertible 4% Liberty Bond $96.33
One @ $50 face value Fourth 4¼% Liberty Bond $48.56
One @ $50 face value Victory 4¾% Note $50.02

The total appraised value came to $314.91.

Also left behind was a photometric survey of the southern sky, and a study of the light curves of novae, including, as a Harvard annual report later put it, "the famous new star of 1918," which had flared in the constellation Aquila. And she was not quite done with another round of revisions to her magnum opus, the North Polar Sequence. When the International Astronomical Union held its first general assembly in Rome the following May, the Commission of Stellar Photometry, of which she had been a member, recognized her "great service to astronomy." "She was one of the pioneers in a difficult field of investigation in which she worked with conspicuous success, and it is deeply regretted that she was unable to finish this her last undertaking."

The next year a Harvard administrative report noted, in passing, something that may have mattered to her more: "She had hardly begun work on her extensive program of photographic measures of variable stars."

Shapley gave her desk to Cecilia Payne. He tried to persuade her to take over Leavitt's unfinished projects, but she had other ideas. After completing a celebrated dissertation on the chemical composition of stars, Payne earned Harvard's first doctorate in astronomy and, under her married name, Cecilia Payne-Gaposchkin, went on to become a full professor and chair of the astronomy department. She never got to meet Leavitt, but she was touched by her story and came to believe that she had been done a great wrong.

"I heard it said when I came to Harvard that what really killed Miss Leavitt was Pickering's requirement that she devise a method by which the photographic magnitudes determined with all the Harvard instruments could be reduced to the same photometric system," she wrote years later. "I cannot believe that he made so unrealistic a request." The judgment seems extreme. This was Leavitt's North Polar Sequence, which she had taken such pride in. But it did keep her away from her Cepheids.

Payne could understand, from a managerial perspective, that it made sense to assign the best of the assistants to tasks that, however onerous, had to be done. "But it was also a harsh decision," Payne wrote, "which condemned a brilliant scientist to uncongenial work, and probably set back the study of variable stars for decades."

Four months after the funeral, Annie Cannon found herself on a steamer bound for Peru, for a tour of the Andes and a visit to the observatory in Arequipa. One evening, after immersing herself in the clear southern skies, she made a note in her diary: "Magellanic Cloud (Great) so bright. It always makes me think of poor Henrietta. How she loved the 'Clouds.' "

3

Like Shapley, Heber Curtis had also moved up in the world, becoming director of the Allegheny Observatory near Pittsburgh. His successor at Lick, a young Swede named Knut Lundmark, carried on the tradition of antagonizing Shapley, claiming that he had been able to pick out individual stars in a pinwheeled nebula called Triangulum or M33 (after the number given it in the eighteenth century by the French astronomer Charles Messier). Assuming (1) that these were the brightest stars in the nebula (which is presumably why he could see them) and (2) that they had the same average magnitude as the brightest stars in the Milky Way, he estimated that M33 was more than a million light-years away. It was, in other words, a full-fledged galaxy.

Shapley quickly challenged him in a letter. Why had his paper not mentioned the work of his friend Adriaan van Maanen, who had measured the rotation of that very spiral, showing it must be small and nearby, or else spinning at impossibly high speeds? Lundmark replied diplomatically and Shapley, for now, was mollified. But he couldn't resist firing off one of the sarcastic slights that were becoming his trademark: "Whether or not you care to recognize that [van Maanen's] measures, if real, practically eliminate the 'island universe' hypothesis, which you seem to espouse at present time more strongly than any one, is not a matter I can properly concern myself about."

Lundmark was not so easily defused, as Shapley learned to his annoyance when he picked up an astronomical journal a few months later and read a new paper entitled "On the Motions of Spirals." The young astronomer was pushing island universes even more vigorously than before, directly questioning the validity of van Maanen's measurements.

Others had also uncovered problems with the data. The British astronomer James Jeans had studied the van Maanen rotations and found that they violated the known laws of physics. Still he was inclined to believe them (they supported his own theory of how galaxies evolve), explaining away the discrepancies with no less than a proposed modification to Newton's law of gravity.

If Lundmark was right, van Maanen's findings were riddled with inconsistencies. He was not suggesting that the astronomer had been sloppy. Measuring such tiny displacements was maddeningly subtle work and open to interpretation. When an object takes 100,000 years to make a single revolution, as van Maanen had concluded, it is not going to move very much in a single year or a decade, or even in the flash of a human lifetime.

No one understood this better than Lundmark. Months later, when he was remeasuring the images of M33, he briefly convinced himself that it really was spinning, so rapidly that "the situation seemed to be rather hopeless for the followers of the island universe theory." Shapley, of course, was delighted. But the crisis of confidence quickly passed. Closer scrutiny assured

Lundmark that he had succumbed to an illusion: there was no sign from this great distance that M33 was rotating at all.

4

Back at Mount Wilson, Hubble was training his sights on Andromeda. It was October 4, 1923. After taking a time exposure of the nebula, he saw what he suspected was another nova. The sky was hazy, so he tried again the next night. This time there seemed to be no question. A second photographic plate showed what appeared to be three novae.

In the observatory office down the mountain in Pasadena, he compared his plate with previous ones taken by Shapley and others. The confirming sign of a nova would be a bright spot appearing where none had been before. One of his flares, however, behaved much differently. Over a period of about a month it had brightened, dimmed, and then brightened again. This was a far more important finding than Hubble had expected. He marked the plate "VAR!" and in February wrote to Shapley: "You will be interested to hear that I have found a Cepheid variable in Andromeda." According to the period-luminosity scale that Shapley himself had calibrated—Shapley's curve—the spiral must be a million light-years away.

Hubble went on to boast that, in fact, he had found two Cepheids and nine novae in Andromeda and expected more to come soon. "Altogether next season should be a merry one and will be met with due form and ceremony."

There he was trying to sound like an Oxford don again. Cecilia Payne later recalled being in Shapley's office when the dispatch arrived: "Here is the letter that has destroyed my universe," she remembered his saying, a bit melodramatically. One wonders whether Payne's memory was a little foggy and the visit actually came later, for Shapley's immediate reaction was hardly one of defeat.

"Your letter telling of the crop of novae and of the two variable stars in the direction of the Andromeda nebula is the most entertaining piece of literature I have seen in a long time," he replied a few days later. Continuing in this vein, he argued that Cepheids with periods as long as a month, like the one Hubble had used for his distance measurement, were unreliable as standard candles. It was most likely that what he had found was not a Cepheid at all. False Cepheids, Shapley said, were discovered all the time. He could show Hubble some examples, if he'd like, from the Harvard plate collection. It was going to take a lot more to convince Shapley that his map of the universe was wrong.

Hubble kept on looking at the sky. Pointing the 100-inch telescope toward the constellation Sagittarius, he zoomed in on an irregular-shaped patch of light that resembled a smaller, dimmer version of the Magellanic Clouds. It had first been spotted in the mid–1880s through a 5-inch telescope by Edward Emerson Barnard, an amateur astronomer with so keen an eye that he was given a fellowship at Vanderbilt and later hired as a professor at the University of Chicago. Later observations showed that Barnard's discovery (usually called by its New General Catalog number, NGC 6822) was a cluster made up of several smaller nebulae and numerous individual stars.

The question, of course, was whether this conglomeration was part of the Milky Way. During 1923 and 1924 Hubble took some fifty pictures of Barnard's cloud, then compared them with images from earlier years using a blink comparator. As he flipped back and forth between the two plates, variable stars pulsed like traffic lights. He found fifteen of them, concluding that most were Cepheids. According to the period-luminosity scale, what could now almost unequivocally be called Barnard's Galaxy was 700,000 light-years away.

Hubble also found more Cepheids in Andromeda and in its neighbor M33. This time he broke the news to Shapley with the gentleness of someone who knew he had changed astronomy. Allowing that it was premature to draw final conclusions, he noted that "the straws are all pointing in one direction and it will do no harm to begin considering the various possibilities involved."

Shapley understood that this was an understatement. "I do not know whether I am sorry or glad," he replied. "Perhaps both."

In a later paper Hubble remarked that NGC 6822 indeed appeared to be "a curiously faithful copy" of the Magellanic Clouds. The galaxy had the same general shape and structure. It was just smaller and dimmer. If one trusted the Cepheids to measure the distances, the reason became clear. Barnard's Galaxy was farther away. For Hubble this consistency was vindication not just of the Cepheid yardstick but of an even grander principle: "The principle of the uniformity of nature thus seems to rule undisturbed in this remote region of space."

5

With few exceptions, the astronomical world almost immediately recognized that the island universe debate had come to an end. Even Henry Norris Russell realized that he had backed the wrong horse. He urged Hubble to announce his findings at the annual meeting of the American Astronomical Society, to be held jointly in Washington with the American Association for the Advancement of Science. In recent years AAAS, for short, has lost its significance as a place to unveil important new science. The huge annual meetings are devoted primarily to educational sessions, giving scientists and reporters a chance to catch up on developments in various fields. For many scientists and science writers the conference is mostly an opportunity to socialize. But in 1925, the meeting was, as Russell put it in a letter to Hubble, "a splendid forum for a major scientific announcement." Russell also assured his young colleague that he was a natural for the recently established AAAS Thousand Dollar Prize for the most outstanding paper of the previous year. He was exasperated when, after arriving in Washington for the meeting, no paper from Hubble had been submitted. It arrived, however, at the last moment, and Russell himself read it from the floor. (In the end Hubble shared the prize with the author of two papers on protozoa inside the digestive tracts of termites.)

An abstract of "Cepheids in Spiral Nebulae" was published in May 1925, more than a year after Hubble had first broken the news to Shapley. The reason for the delay, Hubble told colleagues, was the flat contradiction between his results and van Maanen's. Rechecking his data once again, van Maanen continued to insist that he saw a rotation. But the closer Hubble looked, the more inclined he was to join Lundmark in concluding that the movement was spurious. It was a phenomenon visible to only one man.

To this day no one quite understands where van Maanen went wrong. Perhaps the most compelling theory is that the images of these stellar pinwheels and whirlpools look as though they *should* be turning (as indeed they are, though many times more slowly). Van Maanen may have been subconsciously influenced by his expectations, seeing what he expected to see.

Why Shapley continued to embrace van Maanen's theory, until it was no longer possible to do so, requires no elaborate analysis. Cecilia Payne heard him explain it years later: "After all, he was my *friend*."

Had Shapley stayed at Mount Wilson, booking time on the mighty 100-inch scope, this new grander universe might have been his own. People would turn on their televisions decades later and admire the glorious photos taken by the orbiting Shapley Space Telescope. Instead he is primarily remembered, a little unfairly, as a great astronomer who couldn't see beyond the galaxy, who convinced himself that there could be nothing but the Milky Way.

In an interview decades later, he claimed to have forgotten all about the Great Debate. But as he looked back to the 1920s, the details, a bit scrambled, slowly emerged. His most vivid memory was the false one of Einstein's being there. Shapley said he was surprised that historians were now making so big a deal about the event, contending, a bit disingenuously, that on the "assigned subject matter"—the scale of the universe—he was the clear winner. "I was right and Curtis was wrong on the main point—the scale, the size. It is a big universe, and he viewed it as a small one."

But that seems in retrospect a minor point compared with what Curtis got right—that our galaxy, no matter how large or small, is one among a multitude, a small outpost in what Hubble would come to call "the realm of the nebulae." His protégé Allan Sandage later put it like this: "What are galaxies? No one knew before 1900. Very few people knew in 1920. All astronomers knew after 1924."

CHAPTER 8

The Mysterious K

Youth Who Left Ozark Mountains to Study Stars
Causes Einstein to Change His Mind.

—*Springfield Daily News, February 5, 1931*

In 1925, with the Roaring Twenties half over, John T. Scopes was convicted of violating a Tennessee law against teaching evolution, or any theory denying that the universe had been created as described in the Book of Genesis. Biblical fundamentalists, offended by the suggestion that they were related to monkeys, might have been even more disturbed had they known about recent developments in astronomy. After more than 2,000 years of measuring, scientists were finding little reason to maintain the belief that there was anything special about the position of the sun within the Milky Way, or of the Milky Way itself within the endless sea of galaxies called the universe.

Had there been, say, a Hubble Star Trial resembling the Scopes "Monkey Trial," one can imagine how the prosecution might have argued its case against the propounder of so great a heresy. It would have been difficult, and very unconvincing, to challenge the simple rules of geometry used to triangulate distances within the solar system or even to the nearest stars. Whether the baseline was the diameter of the Earth or the diameter of its orbit around the sun, the reasoning behind the measurements appealed to simple trigonometry and old-fashioned common sense.

More suspect, from the point of view of the astronomical fundamentalists, might have been some of the more indirect techniques—"mathematical mumbo jumbo," a William Jennings Bryan might have objected. And he might have really gone to town with the Cepheids themselves. Perhaps the intrinsic brightness of the variable stars twinkling in the Magellanic Clouds is indeed signaled by how fast they pulse. But is it not a leap of faith to assume that the very same rule applies to Cepheids throughout the universe? Could not the good Lord have made his stars blink in any way he liked?

Here a Clarence Darrow, rising to the defense, might have prodded his client Hubble to remind the jurors how the distances derived from Cepheids harmonized so nicely with measurements made using other techniques, like the average luminosity of a galaxy's brightest stars. Completely independent gauges all seemed to point toward the same conclusion. "But," Bryan might have countered, "do not *all* these methods assume that stars near the Earth are fundamentally the same as stars in the far reaches of the heavens? Is that not a leap of faith?"

Darrow would have been smart to concede the point. What astronomers were taking on faith was the principle of uniformity—that the laws of physics apply equally in all parts of the cosmic realm. It would be an insult to the Creator to think he would design a universe in any other way.

Taking this idea to heart, one could now estimate the distance to any galaxy for which it was possible to pick out individual stars. Assuming that they were similar in kind to those nearby, you could guess their intrinsic brightness and use them as standard candles. Reaching into the astronomical grab bag of variables, novae, and so forth, you could piece together a map of the universe, or at least those regions closer in.

Beyond a certain distance, the method began to falter. Even with Mount Wilson's 100-inch telescope, most nebulae by far were featureless blurs. There was little hope of finding a blinking Cepheid or even an exploding star. For very rough measurements, you could choose a closer galaxy whose distance you had already established and take the whole thing as a standard candle: estimate its intrinsic brightness and assume that the farther galaxies produce approximately the same amount of light. Then the inverse square law would kick in. The method was similar to that used in the eighteenth century when astronomers assumed, both out of ignorance and for convenience, that all stars were equally bright, making dimness a simple gauge of distance—and it was just as prone to error. Maybe the galaxy you were using as your standard was atypical, far brighter or dimmer than most. Still, if you averaged together the brightness of several known galaxies and used that as your yardstick, you could at least make a plausible argument. With this crude method, astronomers were reaching beyond Andromeda, finding galaxies whose light took millions of years to reach their telescopes.

The process was a bit like constructing a tall building with each level resting on the one below. With the diameter of the Earth as a baseline, you measure the distance to the sun. Assuming that figure is correct, you know the width of Earth's orbit, a larger baseline from which to triangulate the very closest stars. Building on this information, you chart the sun's own motion through the galaxy, providing an enormous baseline that, with some statistical sleight of hand, lets you measure the Cepheids and calibrate Henrietta Leavitt's yardstick, and from there you make the next great leap.

The higher you climb, the more precarious the structure becomes. Perched with their telescopes at the loftiest level, astronomers knew it was foolish to be overly confident. At any moment a lower support might buckle. Everything they had built could come tumbling down.

2

Compared with the rickety nature of the distance scales, the celestial speed stick was fairly robust. Because of the Doppler effect, anything that emitted light (stars, galaxies, clouds of gas) could be clocked according to its color shift—a shrieking high-pitched blue if it was speeding toward you, a moaning low-pitched red if it was speeding away.

More specifically, the velocity was determined by the displacement of an object's spectral lines. The method depended on a discovery by the German scientists Gustav Kirchhoff and Robert Bunsen, who found in the 1850s that they could identify chemical elements by burning them in a flame and refracting the glow through a prism. Certain colors would stand out in the spectrum—a combination of bright vertical lines as unique as a fingerprint. Sodium, for example, burns yellow. Seen through a spectroscope, it can be identified by a pair of bright "emission lines" at precise points in the yellow part of the spectrum.

When Kirchhoff and Bunsen made the discovery, the existence of atoms was still controversial. Once they were discovered, the effect could be simply understood: when an atom is energized, its electrons jump into higher orbits. When they fall back down they emit various frequencies of light. Every kind of atom is built a little differently, its electrons arrayed in a specific way, resulting in a characteristic pattern.

For similar reasons, if you shine a light through a gaseous substance, like hydrogen or helium, certain colors will be filtered out. The result in this case is a characteristic pattern of black "absorption" lines interrupting the spectrum—another unique chemical fingerprint. (The same colors marked by the absorption lines would appear as bright emission lines if the element was burned.) A scientist named Joseph von Fraunhofer had shown that lines like these appear in the spectrum of the sun. Using nothing more than a prism, one could stand on Earth and decipher the composition of the glowing orb 93 million miles away.

The natural next step was to add prisms to telescopes and analyze the chemical composition of stars and nebulae. They too exhibited the dark Fraunhofer lines, but not in the expected positions. They were displaced toward the red or blue end of the spectrum. Assuming this was caused by the Doppler effect, you could precisely gauge a galaxy's velocity. A few, like Andromeda, were blue-shifted, approaching the Milky Way, but these were exceptions. Most were severely red-shifted, hurtling away at extremely high speeds.

As the 1920s drew toward a close, astronomers were finding hints of an even stranger phenomenon: the smaller, fainter, and thus presumably more distant a nebula was from Earth, the greater its redshift. Some astronomers dismissed this as an anomaly, a flaw in their techniques. Farther objects were obviously harder to analyze than closer ones. Some kind of systematic error might cause observers to overestimate the more distant redshifts. To correct for the mistake, they added a fudge factor to their calculations, a term they labeled K. It was considered, at this point, no more than a Band-Aid on the equations. When observations improved, it could be peeled off and thrown away.

There was also a more interesting possibility: that the K term was describing a real physical effect—that redshift actually did increase with distance. Grasping for an explanation, some theorists proposed that they were witnessing a heretofore unknown quality of light: the farther it traveled, the more its waves stretched and sagged toward the red end of the spectrum. This became known as the "tired light" theory. Perhaps, some proposed, the cause was some Einsteinian peculiarity of curved space-time.

Finally, it was possible that distant galaxies truly were flying away faster than closer ones. This seemed almost too good to be true. In a universe like this, redshift would provide the ultimate measuring stick. The distance to anything, no matter how far, could be measured as long as you could gather its light. Assuming that all stars are made from the same basic ingredients—hydrogen, helium, et cetera—they can be expected to exhibit the familiar patterns of spectral lines. The more the lines appear to be displaced, the faster the galaxy is moving, and—if the theory is correct—the farther away it is.

During a tour of Europe in 1928, Edwin Hubble, now the toast of his profession, heard reports of this curious redshift-distance connection. When he returned, he decided to investigate. He asked an assistant, Milton Humason, to train the 100-inch Mount Wilson telescope on some distant nebulae and see how their spectra behaved.

Even more than Henrietta Leavitt, Humason seemed an unlikely candidate for the astronomical hall of fame. He began his career at Mount Wilson as a mule driver, carrying material and supplies up the mountain. He married the daughter of the observatory engineer and talked his way into a job as janitor. Given the chance to learn how to make photographic plates, he quickly proved himself to be a very good photographer of starlight and was promoted to assistant astronomer. He had a grade school education.

Humason began his assignment by targeting a nebula so far away that no one had been able to measure its redshift: NGC 7619. The light, funneled through a prism, left its rainbow on the photographic plate. When it was developed it showed two familiar dark lines, indicating the presence of the element calcium. As expected, the lines were pushed toward the red. What was unexpected was how very big this displacement seemed to be. Humason took another picture to confirm the result. Working with a Mount Wilson computer (referred to in his paper simply as Miss MacCormack), he reported that he had clocked a galaxy speeding away at a rate "twice as large as any hitherto observed": 3,779 kilometers per second, or more than 8 million miles per hour, a velocity that would get you from here to the moon in under two minutes.

In the following weeks, Humason measured more redshifts while Hubble scrutinized the results. By now he had compiled a list giving the recessional velocities of forty-six nebulae. He believed he had reliable distances to about half of them, derived from Cepheids, novae, and other yardsticks. For this sample, speed indeed seemed to increase with distance, and in a delightfully straightforward way. Some astronomers had suggested that the relationship might be "quadratic": velocity would increase as the square of the distance. Others suggested more elaborate equations. What Hubble found could hardly have been simpler: A nebula twice as distant as another would be traveling at twice the speed. Triple the distance and the velocity would triple as well. The relationship was what a mathematician calls linear. Take any nebula in the universe and divide its speed by its distance. The result is always the same number—about 150 by Hubble's reckoning.

This powerful number was none other than the mysterious K astronomers had been puzzling over. The "error" was apparently not an error at all but a factor describing how redshift increased with distance. A galaxy that was 1 million light-years from Earth was receding at about 150 kilometers per second. A galaxy 10 million light-years away traveled at 1,500 kilometers per second. Velocity equals distance times K. Or, more significantly, distance equals velocity divided by K. Applying his new formula to NGC 7619, the galaxy Humason had clocked at the breakneck speed of 3,779 kilometers per second, put it at more than 20 million light-years from earth.

It is impossible to sense from Hubble's typically understated paper, or from the droning account he delivered a few years later in a lecture series called "The Realm of the Nebulae," what he felt as the pieces of a new picture of the universe fell into place. Conservative as always, he cross-checked his measurements, testing whether they resulted in absurd conclusions. Using the galaxies' apparent magnitudes and his new Doppler-derived distances, he computed how bright they would really be. The results were reassuring, comfortably within the range of galaxies closer by.

In the following months, Hubble and Humason continued to test the theory. Hubble would calculate a nebula's distance using various measuring sticks and predict the redshift before Humason had even measured it. By now they were clocking galaxies with velocities as high as 20,000 kilometers per second (from Earth to moon in 20 seconds), putting them more than 100 million light-years away.

The numbers were so large that some astronomers were initially doubtful. On a visit to Pasadena, Harlow Shapley told a colleague, "I don't believe these results."

Not that Hubble or his assistant would have cared. Humason remembered one of his last encounters with Shapley, when he was still working on Mount Wilson. Scrutinizing plates of Andromeda with the blink comparator, Humason had spotted what he believed to be stars that varied periodically in brightness. This was more than two years before Hubble made his landmark discovery, establishing with Cepheids that Andromeda was a distant, neighboring galaxy. Humason marked off the places where these anomalies occurred and took the plates to Shapley.

Dismissively explaining why they couldn't be Cepheids, Shapley took out his handkerchief and wiped the plates clean, erasing the data. A few months later he departed for Harvard.

3

At heart, Hubble, like Edward Pickering, was an observer not a theorist, leery of speculating beyond what his eyes could see. It took an Einstein to explain the theory and the mechanism of what astronomers were soon calling the Hubble shift (the K in the equations ceremoniously replaced by an H). Why were the galaxies moving—and why were they, with so few exceptions, all hurtling outward from the Milky Way?

Think of the galaxies as runners in a race. After a certain amount of time has passed, they will be distributed according to their swiftness, the fastest ones farthest from the starting line and the slowest ones closest in. But didn't that imply that there was something special and outrageously non-Copernican about our position in the universe, as the point from which everything else was retreating?

A motionless universe was far easier to fathom, even at first for Einstein. When his own general theory of relativity implied that the cosmos may be expanding, he had rejiggered the equations, committing what he would come to regard as an embarrassing mistake. Now he knew that the adjustment had been superfluous. The universe really moved. Visiting Pasadena in 1931 he told Hubble's wife that her husband's work was "beautiful" and publicly conceded that his earlier conviction of a static universe was wrong. Hubble's hometown newspaper in Missouri picked up the story: "Youth Who Left Ozark Mountains to Study Stars Causes Einstein to Change His Mind."

Einstein was glad he could restore his theory to its pristine form. What his equations now described was a universe in which space itself is expanding. Second by second, the galaxies grow wider apart like dots on an inflating balloon. Viewed this way, Hubble's discovery did not imply that the Earth was in a special position, at the center of the outward rush of everything. From any point in the cosmos the effect would be the same, with galaxies appearing to fly off in every direction. And if you could reverse the clock, everything would become closer and more compact, converging on a single point. The big bang. The universe had a beginning. And maybe it will have an end.

So rarefied a theory, now taken as gospel, was still of secondary interest to most astronomers. Hubble himself was noncommittal about the meaning of the Hubble constant and the Hubble shift. Expanding universe, tired light—it didn't matter. What he knew for certain was that redshift, for whatever reason, increased with distance, and that gave him a way to measure as far as a telescope could see.

CHAPTER 9

The Cosmic Stampede

> The definitive study of the herd instincts of astronomers has yet to be written, but there are times when we resemble nothing so much as a herd of antelope, heads down in tight parallel formation, thundering with firm determination in a particular direction across the plain.
>
> —*J. D. Fernie*

Once he had gotten used to the idea, the enthusiastic Dr. Shapley embraced the new universe with more gusto than the reserved Dr. Hubble would let himself show. Sometimes it almost seemed that Hubble was being willfully perverse. Hewing to tradition, he steadfastly reserved the term "galaxy" for the Milky Way, continuing to refer to Andromeda and the other island universes as nebulae—"extragalactic nebulae," to be exact. Etymologically this might be correct: "galaxy" comes from the Greek word for milk. But Shapley, like everybody else, quickly generalized the term, calling all the island universes galaxies. The bandwagon had almost taken off without him. Now he was back on board.

Though Shapley's old view of the Milky Way as the sole galaxy seemed quainter by the year, he appeared to be right about its enormous size. After all, the same calibration of the Cepheids—Shapley's curve—that had revealed a Milky Way a whopping three hundred thousand light-years from end to end had also been used by Hubble to plumb the distance to Andromeda. From there he had extrapolated outward, using redshift to measure a hundred million light-years into space. If Hubble was right about the size of the universe, it seemed, Shapley must be right about the size of the Milky Way.

And that led to a dilemma. With their new measuring techniques, astronomers could now judge how large other galaxies were. Just take the apparent size and adjust for distance to get the true diameter. Simple enough. But the results of the calculations were disconcerting. None of the galaxies came out to be anywhere near the size of our own. Andromeda measured only a tenth as wide, while the others ranged from a mere thousand light-years to perhaps seventy-five hundred light-years across. The term "island universe" took on a new meaning with the emphasis shifted to the first word. If these distant spirals were islands, Shapley contended, then our own galaxy was a continent.

A few years earlier it would have been acceptable for people to find themselves living in the biggest galaxy around. But perceptions had changed. Shapley had moved the sun from the center of the galaxy, and Hubble had moved the galaxy from the center of the universe. This reversal of perspective was becoming so ingrained that it was a gold standard by which astronomical ideas were judged. If a theory or observation seemed to suggest that we, the observers, happen to occupy an exalted place in the heavens, then it was probably wrong.

Of course it was possible that chance had conspired to put earthlings in a special location. But other discrepancies were harder to dismiss. If the big bang theory was correct, then the size of the universe was an indicator of its age. The larger it was, the longer it had been expanding since the primordial explosion. If the galaxies at the circumference were two billion light-years away, as recent measurements had suggested, it must have taken them two billion years to get there.

Two billion years seemed like a reasonable number. The Earth, however, measured using the technique of radioactive dating, came out to be four billion years old—twice as ancient as the universe that contained it. Something somewhere would have to give.

2

When kinks like these develop in the fabric of knowledge, the fault might lie anywhere in the weave—the result of bad data or a false assumption, the malfunction of a dumb machine or of a human brain. People may see things that turn out to be chimeras. Or miss what is right in front of their telescopes.

At Lick Observatory in California, a Swiss-born astronomer named Robert Trumpler had been studying structures called open clusters in the Milky Way. These stellar aggregations—Pleiades is an example—are smaller and more loosely packed than the globular clusters Shapley used to map the galaxy. Comparing each cluster's true brightness with its apparent luminosity, Trumpler calculated its distance. With this information in hand, he could convert the cluster's apparent diameter into its true diameter—how big it really was.

After measuring a number of them this way, he was forced to a bizarre conclusion: the farther a cluster was from Earth, the larger it appeared to be. What were we doing sitting at the center of so symmetrical an arrangement, surrounded in all directions by increasingly larger star clusters?

More likely, Trumpler reasoned, he was being fooled by an

optical illusion. All his calculations, like those of most every astronomer, took for granted that space was generally transparent, an empty medium through which light could travel unimpeded. If, however, the Milky Way was permeated with a fine cosmic dust, the measurements would be skewed—especially those where the dust was thickest, along the galactic plane. The dimness of a star or galaxy might be due not only to distance but also to this cosmic pollution. The farther the galaxy, the more pronounced the effect. Once Trumpler corrected for the distortion, the clusters turned out to be approximately the same size.

Astronomers had known there was dust in the Milky Way. The surprise was that it could be so pervasive. From almost the beginning, stargazers had contended with light and air pollution here on Earth. As civilization developed and the clarity of the sky degenerated, they placed their observatories higher and higher in the mountains. It hadn't occurred to them that space itself could be so dirty.

And so came the first step toward resolving the Big Galaxy problem, the anomalous size of the Milky Way. When Shapley had taken his measurements, he had been looking through a fog. That made some of his beacons seem much farther than they really were. Once dust was factored into the equations, the home galaxy began to contract in size. That still left it larger than the others, but the adjustment felt like a step in the right direction. More revisions were about to come.

3

While Shapley's map of the galaxy shrunk, the one for the universe grew larger. Again the reason was cosmic dust. Because of the galactic haze, Shapley had underestimated the true brightness of the Milky Way Cepheids, the ones at the foundation of the period-luminosity scale. When Hubble relied on this same standard to measure the distance to Andromeda he was, in effect, mistaking 75-watt bulbs for 60-watt bulbs. If these Cepheids were in fact burning brighter, then we were seeing them across a greater expanse. Using redshift, other galactic distances had been gauged in terms of Andromeda's, so the error rippled outward. Everything was farther than it had appeared.

Astronomers now routinely adjust for the amount of cosmic pollution. Like dust in the atmosphere intensifying the redness of the sunset, dust in interstellar space can be measured by how much it reddens starlight. The lesson, though, took a decade to sink in. For years, one paper after another ignored the dust factor and, errors canceling out other errors, continued to find confirmation for Shapley's original calibration. Looking back years later, the astronomer J. D. Fernie attributed the blindness to a herd instinct: "most of the astronomers of the day simply could not bring themselves to believe that interstellar absorption played any important role."

Dust turned out to be just part of the problem. As far back as the Great Debate, Heber Curtis had suggested that Shapley was overreaching when he assumed that the variables in the Milky Way's globular clusters shared the same relationship between period and brightness as those Henrietta Leavitt had found in the Magellanic Clouds. Both kinds had been lumped together to draw Shapley's curve, the yardstick Hubble had used to measure to Andromeda.

The principle of uniformity encouraged these kinds of generalizations. But were the two variables really the same? If not, the whole distance scale might be askew.

Certain celestial anomalies hinted that this might be true. Even when one allowed for distance, the brightest of the globular clusters in Andromeda—the ones most easily detected—appeared to be inherently dimmer than their counterparts in the Milky Way. A German astronomer named Walter Baade later remembered discussing the discrepancy with Hubble, on cloudy winter nights at Mount Wilson as they waited for the sky to clear. Hubble argued that this might be a case where the principle of uniformity should not be so slavishly followed. After all, he noted, the clusters in the more distant galaxy M33, or Triangulum, were even fainter. Maybe this kind of variation was normal.

Baade had a different idea. Maybe the distance scale was wrong. The clusters in Andromeda and Triangulum were not really emitting less light. They were simply farther than had been reckoned. If so, uniformity would be restored. He soon had a chance to put the hypothesis to a test.

As the mid-1940s approached, many astronomers were off serving in the war. Hubble himself was soon directing ballistic missile tests at the Aberdeen Proving Ground in Maryland. Once Baade, technically an enemy alien, persuaded government authorities that he was not a security threat, he found it easy to book telescope time at Mount Wilson. The periodic blackouts, staged to discourage aerial attacks of Los Angeles, restored the night sky to a primitive blackness. Aiming the 100-inch telescope at Andromeda, he could actually resolve individual stars, not just in the spiral arms but inside the galaxy's dense core.

He discovered what appeared to be two different kinds of starlight. The stars in the galaxy's center and in its globular clusters were colored differently from the "ordinary" stars in the galaxy's outer reaches. That meant the two types must have different chemical makeups. While Leavitt's "classical" Cepheids belonged to what is now called Population I, Shapley's cluster variables belonged to Population II. It seemed a greater stretch than ever to assume that they obeyed the same law relating period and brightness.

When the new 200-inch Hale telescope came on line at Mount Palomar, ninety miles southeast of Mount Wilson, Baade zeroed in on Andromeda for a closer look. The classical Cepheids, he observed, were on average 1.5 magnitudes brighter than the cluster variables. "Instead of one period-luminosity relation," he concluded, "there are actually two."

When the new, brighter value of the classical Cepheids was plugged into the inverse square law, Andromeda turned out to be twice as distant as Hubble had reckoned. And so was the distance to everything else. As the newspapers put it, the universe doubled in size overnight. And, from the perspective of the big bang theory, it doubled in age. It was no longer younger than the Earth.

Baade's discovery completed the explanation of why the other galaxies had seemed so much smaller than our own. That too had been an illusion. If they were farther away, then they were also larger.

Finally, with the new adjustments, the Milky Way was taken down to about 100,000 light-years in diameter—right in between where Curtis and Shapley had put it. It was, in the end, as unremarkable as its stars.

HUBBLE DIED IN 1953. Over the next few years, Allan Sandage, the young astronomer who had served as his last assistant, continued to clock redshifts and make adjustments to the distance scale. For some of his calibrations, Hubble had relied on the brightest star method. Sandage showed that what his old boss had taken as individual stars were actually entire stellar regions. Their intrinsic brightness was therefore much greater, putting the galaxies still farther away. Hubble, as Shapley might have put it, had mistaken trees for asters. The universe was expanding, not just because of the big bang but because of the explosion in astronomical knowledge.

Another way to say it is that the Hubble constant—the number by which you divide a galaxy's velocity to get its distance—was growing smaller and smaller. And so, because of the reciprocal nature of the relationship, the size of the universe continued to grow. Hubble had initially set the constant at 150 kilometers per second per million light-years. More commonly the ratio is expressed using "parsecs" instead of light-years. As the Earth orbits the sun, a star that shows a parallax of 1 arc-second (1/3,600 of a degree) is, by definition, a single parsec (about 3.26 light-years) away. On that scale the Hubble constant had been around 500, Baade knocked it down to 250, and now Sandage to 75. Later he would reduce it again—to 50, ten times smaller than the original value. "The incredible shrinking constant," one astronomer has called it. Every time it gets smaller, the map of the universe grows. So much is packed into that one little number. At its core lie Miss Leavitt's stars.

CHAPTER 10

Ghost Stories

> "What monsters may they be?"
>
> "Impersonal monsters, namely, Immensities. Until a person has thought out the stars and their interspaces, he has hardly learnt that there are things much more terrible than monsters of shape, namely, monsters of magnitude without known shape. Such monsters are the voids and waste places of the sky."
>
> —*Thomas Hardy,* Two on a Tower

For years after Henrietta Swan Leavitt's death in 1921, her presence lived on, not just in her discovery about Cepheid variables but in a ghost story that circulated around Observatory Hill. Cecilia Payne, the young Harvard astronomer (and later department chairman) who had inherited Henrietta's old desk, was amused to hear rumors "that Miss Leavitt's lamp was still to be seen burning in the night, that her spirit still haunted the plate stacks." More likely, she concluded, someone had seen Payne herself as she burned the midnight oil, sometimes working on theories about variable stars.

There is something almost ghostlike in the scant traces Leavitt left in the public record—biographical ectoplasm to be shaped according to one's need. After years of so little recognition, there has followed an almost reflexive rush to mythologize her. A planetarium has been named after her, albeit a virtual one residing only on the World Wide Web. While Harlow Shapley had an entire cluster of galaxies named in his honor, Leavitt (and Annie Cannon) got a crater on the moon.

In brief hagiographies scattered across the Web, the same few scraps of data about Leavitt's life, and often the same sentences, are repeated again and again, all traceable to a few stable sources. She has been turned into a standard bearer by people less interested in her astronomy than in the fact that she was a woman, and deaf. She has been included, absurdly, in a roster of "The World's Greatest Creation Scientists," for no apparent reason other than that she believed in God. She probably would be appalled.

Among the factoids that ricochet through the infosphere is that she was nominated for a Nobel Prize. What happened is that in 1925 Gösta Mittag-Leffler, an elderly Swedish mathematician, heard something about Leavitt's work from a colleague and was impressed enough to write her a letter. He didn't know that she had died.

"Honoured Miss Leavitt," he began. "What my friend and colleague Professor von Zeipel of Uppsala has told me about your admirable discovery of the empirical law touching the connection between magnitude and period-length for the S. Cephei-variables of the Little Magellan's cloud, has impressed me so deeply that I feel seriously inclined to nominate you to the Nobel prize in physics for 1926, although I must confess that my knowledge of the matter is as yet rather incomplete." His expertise was in analytic function theory, not astronomy.

He asked for more information and vowed to handle the matter "with the greatest discretion and in the way that seems to me most likely to further my plan." He also promised to send, for her perusal, a treatise he had written on Sonja

Kowalewsky, a stunning young Russian mathematician, and her correspondence with Karl Weierstrass, the older mentor who had furthered her career. Maybe Mittag-Leffler was hoping Henrietta would become his Sonja.

When the letter arrived at the observatory, it was forwarded to the director, Harlow Shapley. It is hard to know quite what to make of his reply:

"Miss Leavitt's work on the variable stars in the Magellanic Clouds, which led to the discovery of the relation between period and apparent magnitude, has afforded us a very powerful tool in measuring great stellar distances."

Led to the discovery of the relation between period and luminosity? His phrasing suggests that he would deny her credit for her one breakthrough, relegating her back to the role of human computer, the diligent manipulator of data.

The next sentence continues in this vein—faint praise weighed with subtle condescension:

"To me personally [the discovery] has also been of highest service, for it was my privilege to interpret the observation by Miss Leavitt, place it on a basis of absolute brightness, and extending it to the variables of the globular clusters, use it in my measures of the Milky Way."

It is clear whom Shapley would like to nominate for the prize.

UNTIL THE END, Leavitt's title continued to be "assistant." Although Solon Bailey, in his history of Harvard College Observatory, does give her credit for the period-luminosity relationship, he describes the work only in passing and notes, a little belittlingly, "The number of variables included in Miss Leavitt's discussion was unfortunately rather small, but the data have been much increased since that time, especially by the studies of Shapley."

Leavitt probably would have been surprised by how much fuss would later be made over her delightfully simple observation, and by how far Shapley and then Hubble were able to go with it. Given the opportunity—better health, better times—maybe she would have joined them at the forefront. Or maybe not. Barring the discovery of a lost cache of letters, we may never know.

In time, someone else would have discovered Henrietta's law. It is the discovery not the discoverer that matters. Miss Leavitt may have understood this in a way that a Shapley or a Hubble never could. She seemed content to be a small part of a greater thing called science.

In January 1920, the year before she died, a census taker encountered her for the last time in the apartment with her mother on Linnean Street. Among their neighbors were teachers, a salesman for a candy company, a bank clerk, an auditor. Asked to state her occupation, Miss Leavitt replied, honestly and perhaps a bit defiantly, "Astronomer."

2

On a spring afternoon in 1996, seventy-six years to the day since the Great Debate, astronomers gathered at the same lecture hall in Washington for a presentation called, once again, "The Scale of the Universe." The cosmos was seventy-six years older and, if you believed the big bang theory, seventy-six light-years larger in every direction than it had been in 1920.

It was a real debate this time, with opening arguments, rebuttals, and closing statements presented by two celebrated astronomers, Gustav A. Tammann and Sidney van den Bergh. While Tammann argued for a Hubble constant of around 55, van den Bergh put it at about 80. Plugged into the equations of universal expansion, that narrow difference would translate into a universe ranging, depending upon other factors, somewhere between 10 billion and 15 billion light-years in radius.

Large as it seemed, the spread had narrowed in recent years. The lowest values for the constant, around 50, continued to be championed by Allan Sandage, who had taken over where Hubble left off, an opportunity and a burden he once compared to having been Dante's assistant and inheriting *The Divine Comedy*. (The legacy included the incomplete—barely begun, in fact—*Hubble Atlas of Galaxies.*) Then just as the number seemed set in stone, or at least wet cement, an astronomer named Gerard de Vaucouleurs came along and doubled it back to 100, cutting the size of the universe in half. The ensuing controversy became known as the "Hubble Wars."

As Sandage's collaborator and protégé, Tammann was an ideal person to carry on the fight for an older, larger universe, and van den Bergh showed himself to be a formidable opponent. As the debate unfolded, each man in turn challenged the other's choice of standard candles and the manner in which he interpreted them. Watching from the auditorium, members of the audience wore colorful buttons, "Hubble Meters," on which they displayed their own guesses about the constant's value. One might have come away from the debate with the impression that the size of this most fundamental parameter was a matter of opinion.

For all the constant's inconstancy, Harlow Shapley and Heber Curtis both would have been impressed by how far the craft of intergalactic measuring had come. Edward Pickering and Henrietta Leavitt would have been astonished. What is now the world's largest telescope, the Keck, perched atop the 13,800-foot Mauna Kea volcano in Hawaii, collects light with a mirror 10 meters wide. That's 394 inches, or almost twice the size of the 200-inch telescope at Mount Palomar, which was twice the size of the one that Hubble had used to show that Andromeda is really a galaxy. Just when it seemed that mirrors had become as large as physically possible, on the verge of buckling under their own weight, computer and robotics technology stretched the limits further. The mirror on the Keck telescope was made from thirty-six hexagonal sections nudged back and forth by precision pistons so that the whole thing acts like one enormous reflector. Its shape can be constantly adjusted, a few nanometers at a time, to compensate for atmospheric distortion, a technique called adaptive optics. The glass molds itself to the sky.

In fact, by the time of the second debate, there were two Keck telescopes sitting side by side on the mountaintop soon to be linked by an optical interferometer, a computerized device that would combine light from both mirrors, along with that from several smaller scopes, into a single image. The result would be as powerful as a telescope with a mirror 85 meters, or 3,346 inches, across.

For even more acute observations, the Hubble Space Telescope, launched in 1990, was orbiting 380 miles above the Earth, electronically beaming back pictures of deepest space. Among its tasks was looking for Cepheids.

According to the picture that has emerged from these investigations, Andromeda, two million light-years away, now appears to be twice the size of the Milky Way. Both these nebulae mark the far edges of the constellation of galaxies called the Local Group, which also includes Triangulum, the Magellanic Clouds, and several dozen dwarf galaxies.

Nearby are other groups called Sculptor, Maffei, Canes I, Canes II, Dorado . . . so many (more than 150) that most are just given numbers. In addition to the groups are larger "clusters" like Fornax, with 49 galaxies, and Eridanus with 34. Largest of all in this tiny corner of the universe is the Virgo Cluster, which includes another 200 galaxies. Put all these together and you have the Virgo Supercluster—a galaxy of galaxies, numbering in the thousands, spanning 200 million light-years. One of them is the Milky Way.

Farther beyond are the neighboring superclusters: Coma, Centaurus, Hydra, Pavo-Indus, Capricornis, Horologium, Shapley, Sextans—80 of them within a billion light-years. As might be expected, our own Virgo Supercluster turns out to be on the smallish side.

Altogether there are believed to be tens of millions of galaxies just within a billion light-year radius of our solar system—more galaxies than there once were stars.

For all the technological progress astronomy has made, the basic approach to measurement has remained fundamentally the same: use redshift to gauge the recessional velocity of a galaxy or a galactic cluster, then divide that number by the Hubble constant to get the distance.

To calibrate the Hubble scale, Cepheids are still the standard of choice, when you can find them. The Hubble Space Telescope has been spotting them in distant clusters that once remained beyond reach. By the end of its mission in 1993, the European Space Agency's High-Precision Parallax Collecting Satellite, or Hipparcos, had measured parallactic shifts as tiny as one milli-arc-second—1⁄1,000 of one second of one degree.

The thousands of stars whose distances it gauged included a number of Cepheids.

It would be comforting to report that Hipparcos had been able to directly measure the trigonometric parallax of at least one Cepheid in as straightforward a manner as Hipparchus himself had measured the moon. The whole celestial distance scale, the ladders piled on top of ladders, would stand on firmer ground. But the nearest Cepheids are still too distant for even the orbiting satellite to fix any one of them very accurately. A statistical analysis of the better measurements suggested to some astronomers that the whole cosmic distance scale might have to be corrected by 10 percent.

The interpretation is open to dispute. It is probably inevitable that the more astronomers study Miss Leavitt's stars, the less simple they appear. The pulsations of some, including Polaris, include subtle "overtones," secondary rhythms that can throw off the beat. Debates periodically erupt over whether Cepheids of various colors and chemical content have significantly different period-luminosity curves.

In tweaking the Hubble constant, astronomers also now rely on other kinds of pulsating stars, like the RR Lyraes (Shapley's old "cluster" variables) and the Miras. In addition are a wide assortment of secondary candles, those of lesser reliability that are calibrated using Cepheids for measuring beyond where it is possible to pick out individual stars.

During the 1920 debate, one of the arguments against the existence of island universes had been the extreme brightness of a certain nova in Andromeda. Unless the nebula was nearby, the nova would have to be unusually powerful, way off the scale. By the time of the 1996 debate, astronomers spoke comfortably of "supernovae," intense bursts of light coming from exploding stars. A variety of supernovae called Type Ia has been calibrated for use as standard candles. Because of their extreme brightness they have been detected billions of light-years away.

When whole galaxies must be used as standard candles, astronomers can draw on something called the Tully-Fisher method: the larger a spiral galaxy, the faster it will spin. Bigger galaxies are also brighter, so their intrinsic luminosity can be estimated from their rotational speed, which is gauged by measuring Doppler shifts.

The details of this and other methods can quickly become esoteric, but the underlying idea is the same: if you can devise a theory that relates an observable characteristic—of a star, a supernova, a galaxy, a cluster—to its inherent brightness, then you can use it as a standard candle.

The measurements remain fraught with uncertainty. In addition to the universe's outward expansion, galactic clusters are also pulled gravitationally toward one another. These "peculiar motions" result in kinks in the Hubble expansion that must be corrected for. Our Local Group is believed to be gradually falling into the massive Virgo Cluster, a phenomenon called virgocentric flow.

Astronomers must also guard against selection effects, giving too much weight in their calculations to the stars, galaxies, and clusters that are easiest to see. The most well-known of these is the so-called Malmquist bias: The stars you can pick out in a cluster are necessarily the brightest. If you rely on them to compute the average luminosity, the answer will be skewed.

Even with so many ways to go wrong, astronomers have been moving toward a consensus that the universe is a little less than 14 billion years old. Looking out from any vantage point, an observer will be at the center of a bubble extending that many light-years in every direction. It is still a hard idea to get used to. No one is at the center yet everyone is. Wherever you stand, you can see no farther than light has been able to travel since the big bang, the explosion that occurred everywhere and nowhere, that created space and time.

EPILOGUE

Fire on the Mountain

This book began with a parable about a village that learned how to measure all the way to a far-off mountain. Because of a mistaken assumption—that the vegetation on the mountaintop was the same as that on the valley floor—the inhabitants underestimated the distance, only learning of their mistake after they sent an expedition there.

There is a coda to the story. Much later in their history, the villagers established a scientific outpost on the mountain. They built high towers and learned to concentrate light with tubes outfitted with curved lenses and mirrors. Looking out at the unknown expanse beyond them, they realized that their explorations had barely begun. Their mountain was merely a hillock. What lay on the new horizon was a peak that, magnified many times, was of breathtaking grandeur.

It too was fringed with green, and this time, to avoid fooling themselves, the villagers used the *average* height of their own vegetation as a standard yardstick. No more mistaking asters for trees. While the mountain they were standing on was a thousand canyon widths from the village, this new mountain appeared to be approximately a thousand times farther still. This place, they knew, would not be visited in their lifetime, and probably not within the lifetime of their people.

One night up in the tower, one of the scientists saw a brilliant light on the horizon. The remote mountain had exploded in flames. Measuring the intensity of the light, the scientist did some calculations. From the distance to the mountain and its apparent size, he had already estimated how big it was. Now he calculated how much light would be produced if the mountain had caught on fire.

The answer didn't make sense. The flames were so brilliant that they would have to be far more intense than anything resembling ordinary combustion.

When he reported his finding to his colleagues back in the village, they offered several hypotheses. One proposed that some peculiarity of the air may have magnified the light, acting like a natural lens, but few thought that was plausible. More popular was the theory that fire in the distant land burned much hotter, that they had discovered a new kind of energy.

The scientist who had made the observation had a different idea: that this time they had somehow *over*estimated the distance. If the mountain was really a hundred times closer, the anomaly could be explained away. . . .

SOMETHING LIKE THIS happened here on earth. It was 1963 and Maarten Schmidt, a Mount Palomar astronomer, had just ascertained that a starlike ("quasi-stellar") object called 3C273 showed a redshift that would put it several billion light-years from earth, as far as some of the most distant galaxies.

Other quasars were soon found to have even more severe

redshifts. They appeared to be receding from our part of space at a velocity almost as great as light. The Hubble law put them nearly at the edge of the visible universe. For something so remote to shine so brightly, it would have to be emitting the light of thousands of galaxies, the energy generated, perhaps, by matter pouring into the intense gravitational field of a black hole.

Whatever the cause, accepting the immense distance of these fantastic objects caused all kinds of trouble. The quasar 3C273 (the 273rd entry in the *Third Cambridge Catalog of Radio Sources*) expels from its core a jet of light that appears to be traveling at several times the speed of light. That of course would be impossible. Astronomers quickly came up with a more palatable explanation: the superluminal motion is probably an illusion. The jet happens to be coming almost straight at us, making it appear to be much faster than it really is.

There is however another possibility: that 3C273 and all the quasars are really very near by. The velocity of the jet would then be much, much smaller. That also would solve another problem. If the quasars are close to us, then we need not conclude that they are so fantastically bright.

For this to be true, redshift would have to be caused by something other than Hubble expansion—maybe by the old "tired light" theory or some other new physics. If so, the entire universe might be vastly smaller, and there may not have been a big bang.

The idea that the redshifts are "noncosmological" is, to say the least, a minority view. Most astronomers are persuaded by a tightening net of circumstantial evidence that quasars really are blinding beacons lying near the edge of what it is possible to see.

One of the strongest arguments involves a weird phenomenon called gravitational lensing. Sometimes astronomers see two quasars, one right next to the other. The doubling, however, is believed to be an illusion. According to the theory of general relativity, gravity can bend light. Something as massive as a galaxy can act like an enormous piece of curved glass, projecting a double image. If all that is true, then the quasar must be behind the galaxy not in front of it, and therefore very far away.

With each step outward, the act of measurement becomes a little more abstruse. With arithmetic and a ruler you can get from the desk to the window, with trigonometry and a transit you can get to the moon and, with a few assumptions, to the nearest stars.

"With increasing distance, our knowledge fades, and fades rapidly," Hubble once said, in a rare moment of oratorial eloquence. "Eventually, we reach the dim boundary—the utmost limits of our telescopes. There, we measure shadows, and search among ghostly errors of measurements for landmarks that are scarcely more substantial."

Establishing the distance of the quasars requires not only the Hubble law, but the entire framework of Einsteinian relativity. Measuring began as a way to gather data to verify theories. Now the measuring stick itself has become one more theory to test.

Acknowledgments

My increasing curiosity about Henrietta Swan Leavitt could hardly have been sated without the help of Louisa Gilder, who searched the archives at Harvard and Radcliffe with a thoroughness and a connoisseur's eye for detail that went far beyond mere research.

I would also like to thank Kathleen Rawlins and Susan E. Maycock of the Cambridge Historical Commission; Jolene Passey, Faye Leavitt, and Joseph Leavitt of the Western Association of Leavitt Families, and Winston Leavitt of the National Association of Leavitt Families. I greatly benefited from the excellent work of several historians of early-twentieth-century astronomy, including Gale Christianson, J. D. Fernie, Owen Gingerich, Dorrit Hoffleit, Michael Hoskin, Peggy Aldrich Kidwell, Pamela Mack, Robert W. Smith, and Virginia Trimble (their books and papers are cited in my notes). At Harvard College Observatory, Alison Doane guided me through the stacks of photographic plates and old notebooks and helped me find the office where Henrietta Leavitt and the other computers probably worked.

Several people generously read the manuscript, helping me strike a balance between clarity and precision. First I thank the experts: Owen Gingerich, Research Professor of Astronomy and of the History of Science at Harvard University; Alison Doane, Curator of Astronomical Photographs at Harvard College Observatory; her predecessor, Martha Hazen; Stephen Maran of the American Astronomical Society; and Virginia Trimble, Professor of Astronomy and the History of Science at the University of California in Irvine. Just as valuable were the comments of smart, general readers, the kind of people this book is written for: Louisa Gilder, Julie Kinyoun, Douglas Maret, Nancy Maret, and Olga Matlin.

At James Atlas Books and Norton, I would like to thank Mr. Atlas himself, Jesse Cohen, Ed Barber, and Angela Von der Lippe, for their support and enthusiasm. Thanks also go to Trent Duffy, the excellent copy editor, and Esther Newberg and Christine Bauch at International Creative Management.

Notes

Epigraphs

p. ix "Her columns grew longer": Thomas Mallon, *Two Moons* (New York: Pantheon, 2000), p. 11.

p. ix "Then, by means of the instrument at hand": Thomas Hardy, *Two on a Tower* (New York: Harper & Brothers, 1895), p. 33.

Prologue. The Village in the Canyon

p. 1 As readers shall see in chapter 6, my story about the village in the canyon is elaborated from a comment made by Harlow Shapley in the Great Debate of 1920.

p. 6 The view from Tau Ceti is described on pp. 170–71 of Heinlein's *Time for the Stars* (New York: Charles Scribner's Sons, 1956).

Chapter 1. Black Stars, White Nights

p. 9 "We work from morn till night": Jones and Boyd, *Harvard College Observatory*, p. 190. This book and Bailey's *History and Work of the College Observatory* are the two standard sources for the early history of the observatory.

p. 9 Computers earned 10 cents more than a cotton mill worker: Pamela Etter Mack, "Women in Astronomy in the United States,

1875–1920" (bachelor's thesis, Harvard University, 1977). I also referred to her chapter, "Straying from Their Orbits: Women in Astronomy in America," in *Women in Science*, edited by Kass-Simon and Farnes, pp. 72–116.

p. 11 an apparatus of leaf springs: Alison Doane, Curator of Astronomical Photographs at the observatory, has told me that her own research cannot confirm this tale. The repository was built in the 1930s.

p. 14 "It is delightful to see the stars brought out": William Cranch Bond, letter to Harvard President Edward Everett, September 22, 1847, quoted in Jones and Boyd, p. 68.

p. 15 The Great Refractor extended the reach to the fourteenth magnitude: One of the telescope's first great discoveries, in 1848 by William Cranch Bond and George Bond, was Saturn's eighth moon, Hyperion, which is between the fourteenth and fifteenth magnitudes.

p. 15 My portrait of Pickering is drawn from Bailey, pp. 243–52, and Jones and Boyd, pp. 178–82.

p. 16 Eventually Harvard measured and cataloged forty-five thousand stars: Jones and Boyd, p. 202. The results were published in 1908 as *The Revised Harvard Photometry* and appeared in volumes 50 and 54 of the *Annals of the Astronomical Observatory of Harvard College*.

p. 17 The saga of the observing station at Arequipa is entertainingly described in Fernie's *Whisper and Vision*, pp. 153–88. Accounts also appear in Bailey and in Jones and Boyd.

p. 18 "A great observatory should be as carefully organized": Pickering, in a June 28, 1906, address to the Harvard Chapter of Phi Beta Kappa, Harvard University Archives.

p. 19 25 cents an hour amounted to the minimum wage: Adjusted for inflation, 25 cents in 1900 would be worth $5.27 in 2003. Source: "The Inflation Calculator," www.westegg.com/inflation.

p. 19 Portraits of Fleming, Cannon, Maury, and other computers appear in Jones and Boyd.

p. 20 The hours and wages of computing are described in Mack, "Women in Astronomy."

p. 20 "He seems to think that no work": Williamina Paton Fleming diary, March 12, 1900, in the Harvard archives. (The journal is part of a project in which staff members and students apparently were asked to keep a diary showing what life was like at the university.)

p. 20 Pickering's handling of Fleming's request for a raise is documented in his own diary entry for the same date, also in the Harvard archives.

p. 21 Pickering's salary is found in Jones and Boyd, p. 182. The description of his typical workday is from his diary.

p. 21 *The Observatory Pinafore*: Jones and Boyd, pp. 189–93. The author of the parody was Winslow Upton.

Chapter 2. Hunting for Variables

p. 23 "My friends say, and I recognize the truth of it": The letter from HSL to Pickering, dated May 13, 1902, is in the Harvard University Archives.

p. 25 The Leavitt family genealogy is from the excellent database maintained by the Western Association of Leavitt Families at www.leavittfamilies.org, and from information provided by the National Association of Leavitt Families, including a private publication, "Descendants of John Leavitt, the Immigrant, Through His Son, Josiah, and Margaret Johnson," by Emily Leavitt Noyes (Tilton, N.H., 1949), pp. 83, 105, 133.

p. 25 The description of the Leavitt household on Warland Street (now Kelly Road) is from U.S. Census documents, the Cambridge Historical Commission, Cambridge city directories, and a visit to the house, which still stands. The only record I found of Roswell's death was the inscription on his gravestone at Cambridge Cemetery.

p. 26 Erasmus Leavitt's steam engine is described in a pamphlet published by the American Society of Mechanical Engineers, "The Leavitt Pumping Engine at Chestnut Hill Station of the Metropolitan District Commission, Boston, Mass.," printed in honor of the occasion of its designation as a National Historical Mechanical Engineering Landmark by the American Society of Mechanical Engineers, December 14, 1973.

p. 26 For HSL at Oberlin I relied on alumni records in the college archives. Her time at Radcliffe is documented in her transcript, college catalogs, and other records in the Radcliffe College Archives. The period between Radcliffe and Harvard is described in a one-page alumni questionnaire she filled out for Oberlin on April 6, 1908.

p. 28 "Miss Leavitt inherited, in a somewhat chastened form": Bailey's obituary of HSL appeared in *Popular Astronomy* 30, no. 4 (April 1922), pp. 197–99. (It was followed by an article, "Shall We Accept Relativity?" by William H. Pickering, Edward's brother.)

p. 30 "an almost religious zeal": Bailey, *History and Work*, p. 264.

p. 30 The letters between HSL and Pickering during her stay in Beloit are in the Harvard archives.

p. 31 The short biography in which HSL is called "extremely deaf" at Radcliffe appears in volume 8 of the *Dictionary of Scientific Biography*, edited by Charles Coulston Gillispie (New York: Charles Scribner's Sons, 1973).

p. 33 The letter sent from the S.S. *Commonwealth* and a note from HSL's brother acknowledging receipt of her paycheck are in the Harvard archives.

Chapter 3. Henrietta's Law

p. 34 "What a variable-star 'fiend' Miss Leavitt is": The letter, dated March 1, 1905, is from Professor Charles Young at Princeton; it is quoted in Jones and Boyd, *Harvard College Observatory*, p. 367.

p. 34 "In no other portion of the heavens": John Herschel is quoted in Shapley's *The Inner Metagalaxy*, p. 42.

p. 36 "Men said to him, in angry letters": Reverend Leavitt's address to the annual meeting of the American Missionary Association was published as "Preaching: The Main Feature in Missionary Work," *The American Missionary* 39, no. 3 (March 1885), pp. 76–79. (He mistakenly refers to the astronomer as James Herschel.)

p. 36 HSL's trip to Europe is documented in a letter from her to Pickering, dated August 4, 1903, in the Harvard University Archives.

p. 37 "an extraordinary number": From H. S. Leavitt, "1777 Variables in the Magellanic Clouds," *Annals of the Astronomical Observatory of Harvard College* 60, no. 4 (1908), pp. 87–108.

p. 37 The *Washington Post* news brief about HSL appeared January 28, 1906, on p. 4.

p. 37 "the northern star of a close pair": Leavitt, "1777 Variables", p. 96.

p. 37 "boarding with Uncle Erasmus": The *Cambridge City Directory* lists her at his address, 33 Garden Street.

p. 38 the Astronomical and Astrophysical Society of America: This became the American Astronomical Society in 1914, after a long and acrimonious struggle in which the newer science of astrophysics tried to avoid being subsumed as a branch of astronomy. See "How Did the AAS Get Its Name?" by Brant L. Sponberg and David H. DeVorkin, on the society's Web site, www.aas.org/~had/name.html. In two letters to Pickering (December 20, 1905, and December 20, 1906), HSL refers to the group simply as the Astrophysical Society.

p. 38 "It is worthy of notice": Leavitt, "1777 Variables," p. 107.

p. 38 "It has not escaped our notice": Watson and Crick's legendary paper, "Molecular Structure of Nucleic Acids," appeared in *Nature* 171 (1953), pp. 737–38.

p. 39 The letters written during HSL's illness in 1908–1910 are in the Harvard archives. The description of the Leavitt household in Beloit is from census records. That year census takers asked whether anyone in a household was deaf, but the record is ambiguously marked, so it is impossible to tell whether Henrietta was put in that category.

p. 42 Reverend Leavitt's estate is described in his will (Commonwealth of Massachusetts, probate court records for Middlesex County). Henrietta's visit home after his death is documented in letters in the Harvard archives.

p. 42 The visit to Des Moines is mentioned in a letter dated July 3, 1911, in the Harvard archives refers to a Mrs. W. G. H. Strong. In June 1901 Henrietta's sister Martha had married a William James Henry Strong, originally of Council Bluffs, Iowa.

p. 43 "A remarkable relation": From Edward C. Pickering, "Periods of Twenty-five Variable Stars in the Small Magellanic Cloud," *Harvard*

College Observatory Circular no. 173 (March 3, 1912). The relationship between period and luminosity is logarithmic.

p. 44 Cepheid yardstick: HSL was clearly aware of the possibilities opened up by her discovery—if the yardstick could be calibrated. As she wrote on the last page of the report, "It is to be hoped, also, that the parallaxes of some variables of this type may be measured."

Chapter 4. Triangles

p. 45 "I had not thought of making the very pretty use": Quoted in Smith, *The Expanding Universe*, p. 72.

p. 45 An authoritative source for the history of astronomical parallax is Albert Van Helden's *Measuring the Universe*. Good popular accounts include Kitty Ferguson's book by the same name and Alan W. Hirshfeld's *Parallax*. I also relied on two fine histories of astronomy, Arthur Koestler's *The Sleepwalkers* and Timothy Ferris's *Coming of Age in the Milky Way*.

p. 48 The astronomer who first measured the distance to Mars was Gian Domenico Cassini, director of the Paris Observatory.

p. 49 The Transit of Venus occurs twice a century—but not every century. The transits of 1874 and 1882 were followed by the pair scheduled for 2004 and 2012.

p. 55 Ejnar Hertzsprung's use of the sun's motion to triangulate some Cepheids was a bit more complicated than I describe. To determine how much of a star's change in position is due to solar parallax you must first account for how much it has moved on its own. This can be done with statistical methods similar to those Shapley used to measure the Milky Way (see chapter 5).

p. 55 Hertzsprung's article appeared in *Astronomische Nachrichten* 196, pp. 201–10.

p. 56 "that the best service he could render": Bailey, *History and Work*, p. 25.

p. 56 HSL's work diary is in the Harvard University Archives.

p. 56 recovering from stomach surgery: HSL's letter to Pickering, dated May 8, 1913, in the Harvard archives.

p. 57 "It is desirable that the standard scale": HSL, "The North Polar

Sequence," *Annals of Harvard College Observatory* 71, no. 3 (1917), p. 230.

Chapter 5. Shapley's Ants

p. 59 "Her discovery of the relation of period to brightness": Letter from Shapley to Pickering, September 24, 1917, Shapley correspondence, Harvard University Archives.

p. 59 "It is much more natural and reasonable": Kant, *Allgemeine Naturgeschichte und Theorie des Himmels*, published in 1755.

p. 59 Good sources on the early-twentieth-century controversy over the nature of nebulae are Smith's *The Expanding Universe* and J. D. Fernie, "The Historical Quest for the Nature of the Spiral Nebulae," *Proceedings of the Astronomical Society of the Pacific* 82 (1970), pp. 1189–1230.

p. 60 "No competent thinker": Quoted in Struve and Zebergs, *Astronomy of the Twentieth Century*, p. 436.

p. 61 "enveloped and beclouded": Smith, p. 21.

p. 61 an intrinsic magnitude of about –8: Astronomers knew this because the Doppler effect gave a direct reading of how fast a nova was expanding in the direction of Earth. Comparing that number with how fast the nova *appeared* to expand revealed the distance, and from the distance one could gauge the intrinsic brightness. Curtis then reversed the procedure, using the hypothesized brightness to estimate the distance of the novae outside the Milky Way.

p. 62 Shapley tells about the ants in his book *Through Rugged Ways to the Stars*, pp. 65–68.

p. 64 "[T]his proposition scarcely needs proof": Harlow Shapley, "On the Nature and Cause of Cepheid Variation," *Astrophysical Journal* 40 (1914), p. 449.

p. 66 Shapley's artful chain of assumptions was developed in nineteen papers each titled "Studies Based on the Colors and Magnitudes in Stellar Clusters"; some of the later ones were coauthored with his wife, Martha, and other co workers. A full list of citations can be accessed through the Bruce Medalist Web page for Shapley:

www.phys-astro.sonoma.edu/BruceMedalists/Shapley. Also see the detailed bibliography in Smith.

p. 66 living alone in a Cambridge rooming house: Actually the *Cambridge City Directory* lists two: 49 Trowbridge Street and then 49 Dana Street.

p. 66 "Does Miss Leavitt know if they have shorter periods": Letter from Shapley to Pickering, August 27, 1917, Shapley correspondence, Harvard archives.

p. 66 "Miss Leavitt is now absent on her vacation": Letter from Pickering to Shapley, September 18, 1917, ibid.

p. 67 "Her discovery of the relation of period to brightness": Letter from Shapley to Pickering, September 24, 1917, ibid.

p. 67 "I believe the most important photometric work": Letter from Shapley to Pickering, August 20, 1918, ibid.

p. 67 "A few days ago I talked with Miss Leavitt": Letter from Pickering to Shapley, September 14, 1918, ibid. Pickering died on February 3, 1919.

p. 68 Van Maanen's research on the rotation of spiral nebulae was centered on M101, M51, and M33.

p. 69 "So the center has shifted": Shapley's letter to Hale, dated January 19, 1918, is quoted in Owen Gingerich, "Shapley's Impact," in the Harlow Shapley Symposium on Globular Cluster Systems in Galaxies, *Proceedings of the 126th International Astronomical Union Symposium*, Cambridge, Mass., August 25–29, 1986 (Dordrecht, Netherlands: Kluwer Academic Publishers, 1988), pp. 23–36.

p. 69 "Man is not such a big chicken": Shapley, *Through Rugged Ways*, p. 60.

Chapter 6. The Late, Great Milky Way

p. 70 "The spectrum of the average spiral nebula": From the slides Curtis used in his 1920 debate with Shapley, Allegheny Observatory Archives, Pittsburgh.

p. 70 The train trip to Washington is described in Shapley, *Through Rugged Ways to the Stars*, pp. 77–78.

p. 70 The wrangling that took place before the debate is vividly described in Michael A. Hoskin, "The Great Debate: What Really Happened,"

Journal for the History of Astronomy 7, pp. 169–82. I also relied on Virginia Trimble's scholarly and entertaining paper "The 1920 Shapley-Curtis Discussion: Background, Issues and Aftermath," *Proceedings of the Astronomical Society of the Pacific* 107 (1995), pp. 1133–44, and on a perceptive analytical account in Smith, *The Expanding Universe*, pp. 77–86, and Struve and Zebergs, *Astronomy of the Twentieth Century*, pp. 416–20, 441–44.

p. 71 "to some region of space": Quoted in Hoskin's paper.

p. 72 "hammer and tongs": ibid.

p. 72 "miserable" Cepheids: the letter to Russell is quoted in Smith, p. 81.

p. 73 "noble human antique": Shapley's memory of the debate is from *Through Rugged Ways*, pp. 78–81. As with his story about Einstein, Shapley may have also misremembered these details. The records of the meeting in the archives of the National Academy of Sciences are not detailed enough to tell.

pp. 73–79 In addition to the works mentioned above by Hoskin, Trimble, and Smith, my account of the debate relies on Shapley's typescript (the original is in the Harvard University Archives) and Curtis's slides (originals at Allegheny Observatory). Both documents are available at antwrp.gsfc.nasa.gov/diamond_jubilee. The debaters repeated and expanded on their positions in formal papers published in the *Bulletin of the National Research Council* 2 (1921), pp. 171–93 and 194–217.

p. 79 "Debate went off fine in Washington": Quoted in Hoskin's paper.

p. 79 "Now I would know how to dodge things": Shapley, *Through Rugged Ways*, p. 79.

p. 80 "the engulfing of a star": Shapley, "Globular Clusters and the Structure of the Galactic System," *Publications of the Astronomical Society of the Pacific* 30 (1918), p. 53.

p. 80 "Suppose that an observer": From Shapley's *Bulletin* paper.

Chapter 7. In the Realm of the Nebulae

p. 82 "One of the few decent things I have done": Shapley, *Through Rugged Ways to the Stars*, p. 91.

p. 82 For biographical details about Shapley, I referred to *Through Rugged*

Ways, as well as the transcript of the oral-history interviews on which the book is based (Charles Weiner and Helen Wright, "Harlow Shapley," American Institute of Physics, Center for the History of Physics, College Park, Md., June 8, 1966). In addition to serving as my main source on Hubble's life, Christianson's *Edwin Hubble* provided a vivid portrait of Shapley (pp. 129–32).

p. 82 "He is the best student I ever had": Jones and Boyd, *Harvard College Observatory*, p. 432, n. 16.

p. 84 "Bah Jove" and "come a cropper": Shapley, *Through Rugged Ways*, p. 57.

p. 84 Hubble's uneasy relationship with Shapley is documented in the books by Christianson and Smith.

p. 85 "the apartment building on Linnean Street": According to the *Cambridge City Directory*, HSL and her mother had moved there by 1919. Cambridge Historical Commission records show that the building (3–5 Linnean Street and called Linnean Hall) was built in 1914. Rents ranged from $30 to $52.50 a month.

p. 85 "enormous importance in the present discussion": Shapley's letter to HSL, May 22, 1920, is in the Harvard University Archives.

p. 85 Shapley had been overestimating: Owen Gingerich has documented the behind-the-scenes wrangling in "How Shapley Came to Harvard or, Snatching the Prize from the Jaws of Debate," *Journal for the History of Astronomy* 19 (1988), pp. 201–7. Also see Hoskin's "The Great Debate" (cited in the notes for chapter 6).

p. 85 "He is much more venturesome": Gingerich, "How Shapley Came to Harvard," p. 203.

p. 86 "Shapley couldn't swing the thing" : Ibid., p. 204.

p. 86 "So young, so clean, so brilliant": Cannon's diary is in the Harvard archives. Details about Cannon and the Shapley era at Harvard Observatory are from Jones and Boyd, Bailey's *History and Work*, and Haramundanis's *Cecilia Payne-Gaposchkin*, an edition of the astronomer's memoirs edited by her daughter.

p. 87 "I always wanted to learn the calculus": Maury said this to Cecilia Payne (Haramundanis, p. 149).

p. 87 "I shall be happy": Quoted in Jones and Boyd, p. 398.

p. 87 "If one could only go on and on": Fleming diary, Harvard archives.

p. 88 "She was a pure observer": Haramundanis, p. 139.

p. 88 "We shall never understand it": HSL quoted in Haramundanis, p. 140.

p. 88 "Pickering chose his staff to work": Ibid., p. 149.

p. 88 "one of the most important women": Shapley, *Through Rugged Ways*, p. 91.

p. 88 "girl-hours" and "kilo-girl-hours": Ibid., p. 94.

p. 88 "Took flowers to Miss Leavitt": Cannon diaries, Harvard archives.

p. 90 The details of HSL's estate are from probate court records for Middlesex County, Massachusetts.

p. 90 What HSL was working on when she died is from the Bailey obituary cited in the notes for chapter 2 and from *Harvard University Reports of the President and the Treasurer of Harvard College, 1922–1923: The Observatory*, p. 244, Harvard archives.

p. 90 "the famous new star of 1918": *Reports of the President, 1922–1923*, p. 244, Harvard archives.

p. 91 "great service to astronomy": *Transactions of the International Astronomical Union* 1 (1922), p. 69.

p. 91 "She had hardly begun work": *Reports of the President, 1921–1922*, p. 208.

p. 91 For the story of Cecilia Payne and HSL's desk, see Haramundanis, p. 153.

p. 91 "I heard it said when I came to Harvard": Ibid., p. 146

p. 92 "Magellanic Cloud (Great) so bright": Cannon diaries, April 20, 1922.

p. 92 The dispute between Lundmark and Shapley is described in Smith, pp. 105–11.

p. 92 "Whether or not you care to recognize": Quoted in Smith, p. 106.

p. 93 "On the Motions of Spirals": Knut Lundmark, *Publications of the Astronomical Society of the Pacific* 34 (1922), pp. 108–15.

p. 93 Jeans's analysis of van Maanen's data is described in Smith, p. 104. Jeans wasn't opposing the island universe theory. Rather, he thought the Milky Way was considerably smaller than Shapley did. If the spirals were of similar size, they would be much closer to Earth, making the rate of their spin far slower.

p. 93 "the situation seemed to be rather hopeless": Quoted in Smith, p. 108.

p. 94 An artful account of Hubble's measurement of Andromeda is in Christianson, pp. 157–62; also see Smith, pp. 111–26.

p. 94 "You will be interested to hear": Quoted in Smith, p. 114. The correspondence between Hubble and Shapley is divided between the Edwin P. Hubble Manuscript Collection at the Huntington Library in San Marino, Calif., and the Harvard University Archives.

p. 94 "Here is the letter that has destroyed my universe": Haramundanis, p. 209.

p. 95 "Your letter telling of the crop of novae": Quoted in Christianson, p. 159.

p. 95 Before he became a professional astronomer, Barnard had earned enough money spotting new comets (a New York philanthropist was paying $200 for each one) to make a down payment on what he called his "comet house." Today he is best known as the namesake and discoverer of Barnard's star.

p. 96 "the straws are all pointing" and "I do not know whether I am sorry": Quoted in Christianson, p. 159.

p. 96 "a curiously faithful copy": Edwin Hubble, "N.G.C. 6822, A Remote Stellar System," *Contributions from the Mount Wilson Observatory* 302 (1925), p. 410.

p. 96 "The principle of the uniformity": Ibid., p. 432.

p. 97 "a splendid forum": Quoted in Christianson, p. 160. The AAS/AAAS meeting is described in the same passage. More details are in *Popular Astronomy* 33, no. 4 (1925), pp. 158–60. An abstract of Hubble's paper, "Cepheids in Spiral Nebulae," appeared in the same issue beginning on p. 252.

p. 97 Hubble shared the prize: The other winner of what is now called the Newcomb Cleveland Prize was L. R. Cleveland, whose papers had been read before the American Society of Zoologists.

p. 98 "After all, he was my *friend*": Quoted in Haramundanis, p. 209.

p. 98 "assigned subject matter" and "I was right": Shapley, *Through Rugged Ways*, p. 79.

p. 98 "the realm of the nebulae": The phrase is used as the title of Hubble's 1936 book, based on his Silliman lectures at Yale.

p. 98 "What are galaxies?": Sandage, *The Hubble Atlas of Galaxies*, p. 1.

Chapter 8. The Mysterious K

p. 99 "Youth Who Left Ozark Mountains": The newspaper headline, which appeared on page 2 of the paper, is quoted in Christianson, *Edwin Hubble*, p. 210.

p. 99 the Scopes "Monkey Trial": The Harvard historian of science Owen Gingerich has in his collection a telegram from Darrow inviting Shapley to testify.

p. 102 "with some statistical sleight of hand": I'm referring here to the technique of statistical parallax, described in chapter 5.

p. 104 For details of Humason's unusual background see Christianson, pp. 185–86. His work with Hubble is described in Christianson, pp. 192–95, and Smith, *The Expanding Universe*, pp. 180–83.

p. 105 "twice as large as any hitherto observed": Milton Humason, "The Large Radial Velocity of N.G.C. 7619," *Proceedings of the National Academy of Sciences* 15, no. 3 (March 15, 1929), pp. 167–68.

p. 106 at about 150 kilometers per second: Hubble actually expressed K as 500 parsecs (one parsec being 3.26 light-years). His early results are reported in his paper, "A Relation Between Distance and Radial Velocity Among Extra-Galactic Nebula," published in the same issue of *PNAS* as Humason's paper (pp. 168–73). This was followed in 1931 by Edwin Hubble and Milton Humason, "The Velocity-Distance Relation Among Extra-Galactic Nebulae," *Astrophysical Journal* 74, no. 43 (1931), pp. 43–80.

p. 107 "I don't believe these results": Quoted in Christianson, p. 198.

p. 107 The encounter between Shapley and Humason is described by Christianson, p. 151. In *The Expanding Universe*, Smith gives good reasons to believe the story may be true (p. 144, n. 122).

p. 108 Einstein's calling Hubble's work "beautiful": See Christianson, p. 211.

Chapter 9. The Cosmic Stampede

p. 109 "The definitive study of the herd instincts of astronomers": J. D. Fernie, "The Period-Luminosity Relation: A Historical Review," *Publi-*

cations of the Astronomical Society of the Pacific* 81, no. 483 (December 1969), pp. 719–20.

p. 109 For a detailed account of the controversy over the Milky Way's seemingly anomalous size, see Smith, *The Expanding Universe*, pp. 153–56.

p. 110 If these distant spirals were islands: Ibid., p. 154. For a while, Shapley toyed with what he called the Super-Galaxy Hypothesis, in which the Milky Way consisted of a confederation of several smaller galaxies—the globular clusters—bunched together.

p. 111 Trumpler reported his discovery of cosmic dust in his paper "Absorption of Light in the Galactic System," *Publications of the Astronomical Society of the Pacific* 42 (1930), pp. 214–27.

p. 113 "most of the astronomers of the day": Fernie, "Period-Luminosity Relation," pp. 716–17.

p. 114 Baade gave a nice personal account of how he recalibrated the Cepheid scale—including a description of his conversation with Hubble—in a talk to the Astronomical Society of the Pacific, later published as "The Period-Luminosity Relation of the Cepheids," *Publications of the Astronomical Society of the Pacific* 68 (1956), pp 5–16. Christianson provides further details in *Edwin Hubble*, pp. 291–93.

p. 115 "Instead of one period-luminosity relation": Baade, p. 11. Along with the cluster variables, he discovered that the Cepheids occasionally found in globular clusters also belonged to Population II.

p. 116 For the revisions to the Hubble constant, see Virginia Trimble, "H_0: The Incredible Shrinking Constant, 1925–1975," *Publications of the Astronomical Society of the Pacific* 108 (December 1996), pp. 1073–82.

Chapter 10. Ghost Stories

p. 117 "What monsters may they be?": The passage appears on p. 34 of Hardy's *Two on a Tower*.

p. 117 "that Miss Leavitt's lamp was still to be seen burning": Haramundanis, *Cecilia Payne-Gaposchkin*, p. 153.

p. 117 Leavitt's scant traces: Annie Cannon was, by contrast, quite a pack rat. The Harvard University Archives are stuffed with diaries, guest books, photo albums, letters—nothing, it seems, was discarded. The scant mention these papers make of Henrietta Leavitt make one wonder whether they were even friends.

p. 117 A virtual planetarium: The Henrietta Leavitt Flat Screen Space Theater is located at www.thespacewriter.com.

p. 118 crater on the moon: Harry Lang, "Six Moon Craters Named for Deaf Scientists," *The World Around You*, January–February, 1996 (published by Gallaudet University, Washington, D.C.).

p. 118 "The World's Greatest Creation Scientists" can be found at www.creationsafaris.com. The criteria for compiling the list are so loose that it also includes Sir Francis Bacon, Johannes Kepler, Leonardo da Vinci, William and John Herschel, and even Galileo.

p. 118 "Honoured Miss Leavitt": Letter from Mittag-Leffler to HSL, February 23, 1925, in Shapley correspondence, Harvard archives.

p. 118 Sonja Kowalewsky: The name also appears in English as Sonya Kovalevskaya. Other variations are Sonja Kowalewski, Sophia Kovalevsky, Sofia Kovalevskaia, and Sofya Kovalevskaya.

p. 119 "Miss Leavitt's work on the variable stars": Letter from Shapley to Mittag-Leffler, March 9, 1925, Shapley correspondence, Harvard archives.

p. 119 "The number of variables included in Miss Leavitt's discussion": Bailey, *History and Work*, p. 185.

p. 120 The 1996 "The Scale of the Universe" debate was amply documented in six papers appearing in *Publications of the Astronomical Society of the Pacific* 108 (December 1996): Jerry T. Bonnell, Robert J. Nemiroff, and Jeffrey J. Goldstein, "The Scale of the Universe Debate in 1996," pp. 1065–67; Owen Gingerich, "The Scale of the Universe: A Curtain Raiser in Four Acts and Four Morals," pp. 1068–72; Virginia Trimble, "H_0: The Incredible Shrinking Constant, 1925–1975," pp. 1073–82; G. A. Tammann, "The Hubble Constant: A Discourse," pp. 1083–90; Sidney van den Bergh, "The Extragalactic Distance Scale," pp. 1091–96; and John N. Bahcall, "Is H_0 Well Defined?" p. 1097.

p. 121 having been Dante's assistant: Christianson, *Edwin Hubble*, p. 363.

p. 121 Besides the value of the Hubble constant, other factors affecting the size of the universe include its shape and the value of a parameter called the cosmological constant.

p. 121 The Hubble Wars are described in Overbye, *Lonely Hearts of the Cosmos*, pp. 263–84.

p. 121 Wonderful maps of clusters and superclusters can be found at www.anzwers.org/free/universe/galaclus.html.

p. 124 corrected by 10 percent: M. W. Feast and R. M. Catchpole, "The Cepheid PL Zero-Point from Hipparcos Trigonometric Parallaxes," *Monthly Notices of the Royal Astronomical Society* 286 (1997), L1–L5. More recently, the biggest news about Hipparcos has been a controversy over the accuracy of its triangulation of the Pleiades: X. Pan, M. Shao, and S. R. Kulkarni, "A Distance of 133–137 Parsecs to the Pleiades Star Cluster," *Nature* 427 (2004), p. 396.

Epilogue: Fire on the Mountain

p. 130 "With increasing distance, our knowledge fades": Hubble, *Realm of the Nebulae*, p. 202.

Selected Bibliography

Bailey, Solon I. *The History and Work of Harvard Observatory, 1839 to 1927.* New York: McGraw-Hill, 1931.

Christianson, Gale E. *Edwin Hubble: Mariner of the Nebulae.* Chicago: University of Chicago Press, 1995.

Evans, David S., Terence J. Deeming, Betty Hall Evans, and Stephen Goldfarb, eds. *Herschel at the Cape: Diaries and Correspondence of Sir John Herschel, 1834–1838.* Austin: University of Texas Press, 1969.

Ferguson, Kitty. *Measuring the Universe: Our Historic Quest to Chart the Horizons of Space and Time.* New York: Walker, 1999.

Fernie, Donald. *The Whisper and the Vision: The Voyages of the Astronomers.* Toronto: Clarke, Irwin, 1976.

Ferris, Timothy. *Coming of Age in the Milky Way.* New York: William Morrow, 1988.

Haramundanis, Katherine, ed. *Cecilia Payne-Gaposchkin: An Autobiography and Other Recollections.* Second ed. Cambridge: Cambridge University Press, 1996.

Hirshfeld, Alan W. *Parallax: The Race to Measure the Cosmos.* New York: W.H. Freeman, 2001.

Hoffleit, Dorrit. *Women in the History of Variable Star Astronomy.* Cambridge, Mass.: American Association of Variable Star Observers, 1993.

Hubble, Edwin. *The Realm of the Nebulae.* New Haven: Yale University Press, 1936.

Jones, Bessie Zaban, and Lyle Gifford Boyd. *The Harvard College Observatory: The First Four Directorships, 1839–1919.* Cambridge, Mass.: Belknap Press, 1971.

Kass-Simon, G., and Patricia Farnes, eds. *Women of Science: Righting the Record.* Bloomington: Indiana University Press, 1993.

Koestler, Arthur. *The Sleepwalkers: A History of Man's Changing Vision of the Universe.* New York: Macmillan, 1959.

Layzer, David. *Constructing the Universe.* New York: Scientific American Library, 1984.

Overbye, Dennis. *Lonely Hearts of the Cosmos: The Scientific Quest for the Secret of the Universe.* New York: HarperCollins, 1991.

Sandage, Allan. *The Hubble Atlas of Galaxies.* Washington, D.C.: Carnegie Institution, 1961.

Shapley, Harlow. *The Inner Metagalaxy.* New Haven: Yale University Press, 1957.

———. *Through Rugged Ways to the Stars.* New York: Charles Scribner's Sons, 1969.

Smith, Robert W. *The Expanding Universe: Astronomy's "Great Debate" 1900–1931.* Cambridge: Cambridge University Press, 1982.

Struve, Otto, and Velta Zebergs. *Astronomy of the Twentieth Century.* New York: Macmillan, 1962.

Van Helden, Albert. *Measuring the Universe: Cosmic Dimensions from Aristarchus to Halley.* Chicago: University of Chicago Press, 1985.

Zeilik, Michael, and John Gaustad. *Astronomy: The Cosmic Perspective.* Second ed. New York: John Wiley & Sons, 1990.

Index

Page numbers in *italics* refer to illustrations.